Synthesis Lectures on Engineering, Science, and Technology

The focus of this series is general topics, and applications about, and for, engineers and scientists on a wide array of applications, methods and advances. Most titles cover subjects such as professional development, education, and study skills, as well as basic introductory undergraduate material and other topics appropriate for a broader and less technical audience.

Baris Burnak · Nikolaos A. Diangelakis ·
Efstratios N. Pistikopoulos

Integrated Process Design and Operational Optimization via Multi-parametric Programming

Second Edition

Springer

Baris Burnak
Texas A&M University
College Station, TX, USA

Nikolaos A. Diangelakis
Technical University of Crete
Chania, Greece

Efstratios N. Pistikopoulos
Texas A&M University
College Station, TX, USA

ISSN 2690-0300 ISSN 2690-0327 (electronic)
Synthesis Lectures on Engineering, Science, and Technology
ISBN 978-3-032-10180-8 ISBN 978-3-032-10181-5 (eBook)
https://doi.org/10.1007/978-3-032-10181-5

This Springer imprint is published by the registered company Springer Nature Switzerland AG
The registered company address is: Gewerbestrasse 11, 6330 Cham, Switzerland

Preface

The book presents a comprehensive optimization-based theory and framework that exploits the synergistic interactions and tradeoffs between process design and operational decisions that span different time scales. Conventional methods in the process industry often isolate decision-making mechanisms with a hierarchical information flow to achieve tractable problems, risking suboptimal, even infeasible operations. The foundations of systematic model and optimization based strategies for simultaneous process design, scheduling, and control optimization are detailed with the aim to achieve reduced cost and energy consumption in process systems, featuring a unified multi-parametric optimization and control perspective. The material covered in this book is well suited for the use of industrial practitioners, academics, and researchers.

In Chap. 1, a historical perspective on the milestones in model-based design optimization techniques is presented along with an overview of the state-of-the-art mathematical tools to solve the resulting complex problems. Chapters 2 and 3 discuss two fundamental concepts that are essential for the reader. These concepts are (i) mixed integer dynamic optimization problems and two algorithms to solve this class of optimization problems, and (ii) developing a model-based multi-parametric programming model predictive control framework. These tools are then used to systematically evaluate the tradeoffs between different time scale decisions on (i) design and control, (ii) scheduling and control, and (iii) integrated design, scheduling, and control problems. We present various illustrative examples on processing units, including continuous stirred tank reactors, distillation columns, and combined heat and power regeneration units, along with discussions of other relevant work in the literature for each class of problems.

The second edition of the book includes updated literature reviews in the areas of advanced control, simultaneous design and control, simultaneous control and scheduling and the integration of operational and design optimization. Furthermore, we highlight the classification of different methodologies for integrating operational and design optimization with special attention to the different scales that need to be considered. Finally, we provide insights regarding the use of AI-based tools in conjunction with first-principle modeling approaches.

Tulsa, USA Baris Burnak
Chania, Greece Nikolaos A. Diangelakis
College Station, USA Efstratios N. Pistikopoulos

Acknowledgments The authors are grateful for the financial support from the National Science Foundation (Grant no. 1705423), the RAPID Institute, Shell Oil Company and the Texas A&M Energy Institute. Many thanks also to Ms. Sahithi Srijana Akundi for her useful comments and input.

Competing Interests The authors have no competing interests to declare that are relevant to the content of this manuscript.

Contents

1 **An Introduction to the Grand Unification of Process Design and Operational Optimization** 1
1.1 Overview 2
1.2 Early Efforts in Design Optimization Under Uncertainty 4
1.3 Integration of Process Control in Design Optimization 7
1.4 Classification of Integration Methodologies 13
1.4.1 Top-Down Integration 13
1.4.2 Bottom-Up Integration 14
1.4.3 Integration of Operability and Control 17
1.5 Towards the Grand Unification of Process Design, Scheduling, and Control 19
1.6 Current Challenges and Future Directions 23
1.6.1 Need for an Industrial Benchmark Problem 23
1.6.2 Robust Advanced Control and Scheduling Strategies 24
1.6.3 Flowsheet Optimization, Process Intensification, and Modular Design Opportunities 25
1.6.4 Integration of Scales 25
1.6.5 Theoretical and Algorithmic Developments in MIDO 32
1.6.6 Software Development 32
1.6.7 First-Principle Modeling in the Age of Artificial Intelligence 32

2 **Mixed-Integer Dynamic Optimization for Simultaneous Process Design and Control** 33
2.1 Overview 33
2.2 Preliminaries 35
2.3 Review of Existing Solution Approaches for MIDO 35
2.4 An Alternative Approach 40
2.4.1 Primal Problem 40
2.4.2 Construction of the Master Problem 41

2.5 New Algorithm I ... 42
2.6 Remarks on MIDO Algorithm I ... 43
2.7 Extending Algorithm I to Handle Time-Dependent Binary Variables ... 46
2.8 Using Outer-Approximation to Construct the Master Problem ... 47
2.9 New MIDO Algorithm II ... 49
2.10 Remarks on Algorithm II ... 49
2.11 Example—Binary Distillation Column ... 50
2.12 Concluding Remarks ... 52

3 PAROC: PARametric Optimisation and Control Framework ... 57
3.1 Overview ... 57
3.2 The PAROC Framework ... 59
3.2.1 High-Fidelity Modeling and Analysis ... 60
3.2.2 Model Approximation ... 60
3.2.3 Multi-parametric Programming ... 61
3.2.4 Multi-parametric Moving Horizon Policies ... 65
3.2.5 Multi-parametric Moving Horizon Estimation ... 68
3.2.6 Remarks ... 70
3.2.7 Software Implementation and Closed-Loop Validation ... 71
3.3 Illustrative Example ... 73
3.4 A Case Study: Residential Combined Heat and Power System ... 75
3.5 Concluding Remarks ... 81

4 Integrating Process Design Optimization and Advanced Model Based Control Strategies ... 83
4.1 Overview ... 83
4.2 Simultaneous Design and Control Optimization via PAROC ... 84
4.2.1 Illustrative Example ... 90
4.3 Concluding Remarks ... 113

5 Process Scheduling and Control via Multi-parametric Programming ... 115
5.1 Overview ... 115
5.2 Problem Definition ... 118
5.3 Problem Formulation ... 118
5.3.1 Incorporating Process Scheduling in the PAROC Framework ... 119
5.4 Examples ... 124
5.4.1 Example 1 Single CSTR with Three Inputs and Three Outputs ... 124
5.4.2 Example 2 Two CSTRs Operating in Parallel ... 135
5.5 Concluding Remarks ... 138

6 Simultaneous Process Design, Scheduling, and Advanced Model Based Control 139
6.1 Overview 139
6.2 Problem Definition 142
6.3 Problem Formulation 142
6.4 Case Studies 145
6.4.1 CSTR with Three Inputs and Three Outputs 145
6.4.2 Two CSTRs Operating in Parallel 147
6.4.3 Residential Combined Heat and Power (CHP) Unit 148
6.4.4 Two CHPs Operating in Parallel 154
6.4.5 Multipurpose Batch Process 157
6.5 Concluding Remarks 166

Appendix A: Supporting Information for Chap. 2 169

Appendix B: Supporting Information for Chap. 4 177

Appendix C: Supporting Information for Chap. 5 187

Appendix D: Supporting Information for Chap. 6 191

References 199

An Introduction to the Grand Unification of Process Design and Operational Optimization

1

The Process Systems Engineering (PSE) community has been accumulating formidable knowledge and know-how on mathematical modeling techniques in the fields of process design and operations, and has developed efficient tools to solve these advanced models since the idea of integrating long and short term systems decisions was first pitched in 1967 [1]. Today, mathematical models and optimization techniques are indispensable tools to explore the underlying mechanics of our systems and to make more educated decisions that improve efficiency, sustainability, and safety of the processes we design and operate. With the advance of operations research techniques and computing power, the problems we tackle have become more inclusive and more complex. It has been long established that the early design problem should be studied simultaneously with the operational time-variant decisions to improve the operability and flexibility of the process under variable internal and external plant conditions, and consequently to achieve more reliable, economically more favorable, and inherently safer processes [2]. The most recent efforts towards simultaneous consideration of design and operational decisions explore effective methodologies to integrate the short term process regulatory decisions (process control) and longer term economical decisions (scheduling) through mixed-integer dynamic optimization (MIDO) formulations. The proposed solution tools and techniques for this class of integrated problems include (i) discretizing the dynamic high fidelity representation of the process through orthogonal collocation on finite elements followed by solving a mixed-integer nonlinear programming problem [3], (ii) "back-off" approach to ensure constraint satisfaction under some assumed worst-case scenario [4–6], and (ii) multi-parametric programming to explicitly represent the operational strategies to derive tractable and equivalent MIDO formulations [7].

In this book, we present a systematic, process agnostic theory and framework to simultaneously account for process design, scheduling, and optimal control problems based on a model based multi-parametric programming approach. This novel methodology has been

B. Burnak et al., *Integrated Process Design and Operational Optimization via Multi-parametric Programming*, Synthesis Lectures on Engineering, Science, and Technology, https://doi.org/10.1007/978-3-032-10181-5_1

developed over the last 20 years using the state-of-the-art techniques in operations research and chemical engineering disciplines. In this first chapter, we present a historical perspective on the milestones in model-based design optimization techniques and the developed tools to solve the resulting complex problems. We examine the progress spanning more than five decades, from the early flexibility analysis and optimal process design under uncertainty to more recent developments on the simultaneous consideration of process design, scheduling, and control. In Chap. 2, we present the concept of mixed-integer dynamic optimization (MIDO) along with a rigorous algorithm to solve this class of problems. MIDO is one of the two fundamental constituents of the framework to integrate process decisions that will be discussed throughout the book. Therefore here, we will explore how to incorporate process design and classical control problems, demonstrated on a well-established binary distillation column model. The second fundamental concept to have an in-depth understanding is model based advanced controller design and the theory of multi-parametric programming, which is discussed in Chap. 3. Model Predictive Control (MPC) has become increasingly more popular over the years due to its strength in imposing dynamic constraints on process variables and has been applied to a plethora of applications in numerous industries [8]. However, due to the implicit nature of the MPC structure, it is not trivial to integrate in an MIDO as discussed in Chap. 2. Hence in Chap. 3, we will introduce the PARametric Optimisation and Control (PAROC) framework, which allows for deriving the exact explicit expressions for an MPC problem. In the following chapters, we will use these concepts in tandem to develop the complete framework to integrate process design, scheduling, and control decisions. In Chap. 4, we will first consider the design and control problems, where the control strategy is an MPC, as opposed to the classical PID control in Chap. 2. In Chap. 5, we will cover the scheduling and control problem, and finally in Chap. 6, we will put all pieces together for the complete framework. We use an illustrative example which will be explored further in each chapter, and we will also provide more complex chemical engineering applications to demonstrate the steps in detail. A summary of the book chapters is demonstrated in Fig. 1.1.

This book is intended for readers with a foundational understanding of mathematical programming and differential-algebraic equations, as well as basic familiarity with industrial practices in process design and optimization. We invite the reader to explore the power of multi-parametric programming as a transformative approach to integrating process design and operational optimization, bridging theory and practice to drive more efficient and intelligent decision-making in complex systems.

1.1 Overview

In this introduction chapter, we present a historical perspective on the development and progress of modern process design techniques that account for the dynamic variability introduced by the process control and scheduling decisions. In retrospect, we observe the evolution of methodologies from fundamental analyses on design and process uncertainty

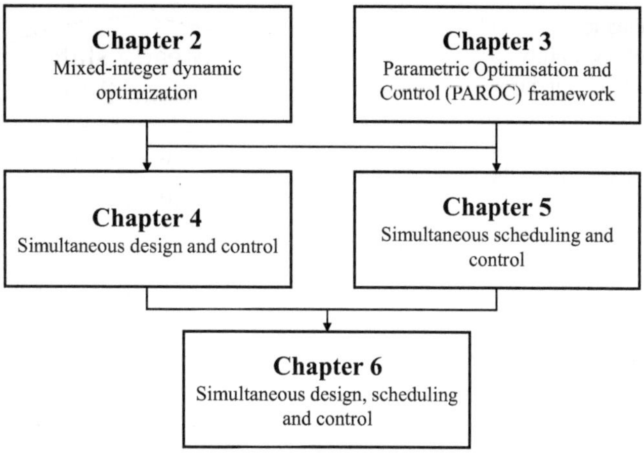

Fig. 1.1 An illustrative summary of the book chapters

at steady state to dynamic complex models that explicitly encapsulate the scheduling and control decisions, as illustrated in Fig. 1.2, and summarized as follows.

i. *Flexibility analysis and flexibility index.* The early stages for design optimization under uncertainty. The studies here analyze the steady state feasibility of a nominal process design under a set of unknown process parameters and unrealized operating decisions, as we will discuss in Sect. 1.2.
ii. *Dynamic resilience and controllability analysis.* Here, the researchers investigate the dynamic response of a system in closed loop, its interdependence with process design, and attempt to develop the "perfect controller" simultaneously the process that the controller can act on. Such attempts will be demonstrated in Sect. 1.3.
iii. *Complete integration of design, control, and operational policies.* The focus of the most recent studies in the field. The goal is to model tractable dynamic design optimization problems that account for the scheduling and control decisions to guarantee the operability and even profitability of the operation under all foreseeable conditions. These approaches will be discussed in Sect. 1.5.

Clearly, it would be inaccurate and redundant trying to reduce down the individual research efforts to a single category. The literature is noticeably diverse in this field with numerous different approaches. However, we find it useful to classify into certain schools of thought that are also in alignment with the historical progress of the field. In Sect. 1.6, we further seek to pose the pivotal questions on future challenges and opportunities for the seamless integration of the design, scheduling, and control problems based on the cumulative knowledge of the PSE community and the current trends in the academia.

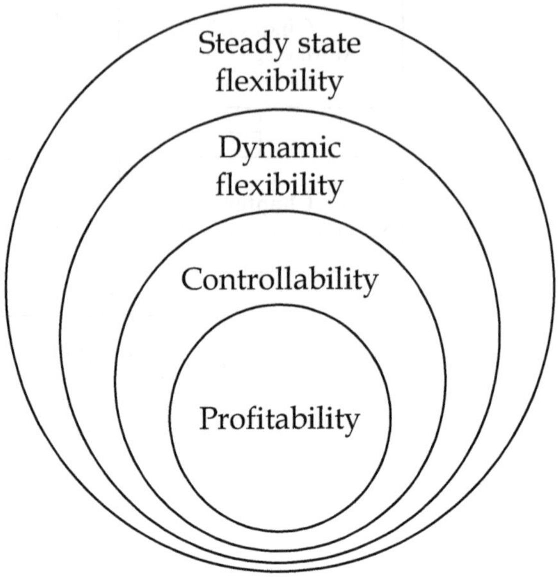

Fig. 1.2 A schematic representation of operability and process economics indices. Note that the historical progress on design optimization approaches has started from the outermost layer and proceeded towards the inner layers with time. [Adapted from [2]]

1.2 Early Efforts in Design Optimization Under Uncertainty

The ongoing collective efforts towards the grand unification of design, scheduling, and control was inaugurated through steady state design under uncertainty in plant conditions. Takamatsu et al. [9] estimated the undesirable effects of variations in system parameters, measured process disturbances, and manipulated variables on plant performance by sensitivity analysis on a linearized model. Nishida et al. [10] adopted the notion of sensitivity analysis to structure a min-max problem for design optimization, presented by Eq. 1.1.

$$\begin{aligned} \min_{des} \max_{\theta} \quad & C(x, des, \theta) \\ s.t. \quad & h(x, des, \theta) = 0 \\ & g(x, des, \theta) \leq 0 \\ & \underline{\theta} \leq \theta \leq \overline{\theta} \end{aligned} \tag{1.1}$$

where x is the vector of states of the system, des is the vector of design variables including the steady state manipulated variables, θ is the vector of parameters that agglomerates the system uncertainties and process disturbances. Equation 1.1 is one of the first notable attempts to systematically assess the trade-off between minimizing the investment cost and improving the flexibility of the process design. However, this strategy yields conservative solutions since it does not distinguish the time-invariant design variables and time-variant manipulated variables. Grossmann and Sargent [11] remedied this issue by treating the time-sensitive variables (i.e. manipulated actions and design variables that can be modified in the future) and fixed design variables separately. They further adopted the parametric optimal design problem proposed by Kwak and Haug [12], and formulated an objective function to minimize the average cost over the expected range of parametric uncertainty, as presented by Eq. 1.2.

$$\begin{aligned} \min_{u,des} \quad & E\{C(x, u, des, \theta)\} \\ s.t. \quad & \max_{\theta \in \Theta} \quad g_i(x, u, des, \theta) \leq 0, \quad i = 1, 2, ..., t \end{aligned} \tag{1.2}$$

where the expected cost function is defined the joint probability distribution of the parameter set θ. Equation 1.2 requires solving infinite nonlinear programming (NLP) problems. Grossmann and Sargent [11] proposed an efficient solution procedure for a special case of Eq. 1.2, where each constraint g_i is monotonic in θ, through discretization of the problem over the parameter space. However, solving the NLP problem at a finite number of θ realizations does not ensure the feasibility of the design. This issue is addressed by Halemane and Grossmann [13] through reformulating an equivalent design feasibility constraint as presented by Eq. 1.3.

$$\max_{\theta \in \Theta} \min_{u \in U} \max_{i \in I} g_i(x, u, des, \theta) \leq 0 \tag{1.3}$$

The max-min-max problem in Eq. 1.3 mathematically expresses the feasibility question "For all the uncertainty realizations Θ, does there exist a control action u such that the constraint set g is feasible?". Equation 1.3 was employed in a multiperiod design optimization problem, where the deterministic uncertain parameter θ was allowed to vary within a prespecified range [13]. The feasibility constraint then laid the foundation for the concept of feasibility index, F, proposed by Swaney and Grossmann [14], as given by Eq. 1.4.

$$\begin{aligned} F = \max \quad & \delta \\ s.t. \quad & \max_{\theta \in \Theta} \min_{u \in U} \max_{i \in I} g_i(x, u, des, \theta) \leq 0 \\ & T(\delta) = \{\theta \mid (\theta^{nom} - \delta\Delta\theta^-) \leq \theta \leq \theta \mid (\theta^{nom} + \delta\Delta\theta^+\} \end{aligned} \tag{1.4}$$

where T is the hyperrectangle for the uncertain parameters, δ is the scaled parameter deviation, and the superscript nom denotes nominal conditions. Equation 1.4 is the first significant attempt to quantify the degree of flexibility of a process design, and has been exploited by numerous studies on design optimization and process operability. However, Eq. 1.4 constitutes a nondifferentiable global optimization problem and is still quite challenging to solve. Therefore, it requires simplifying assumptions and approximations to maintain a tractable problem. Swaney and Grossmann [15] introduced a heuristic vertex search method and an implicit enumeration scheme for the special case where the critical uncertainty realizations are assumed to lie at the vertices of the hyperrectangle $T(\delta)$. Clearly, this assumption fails to hold when the feasible space of the design problem is non-convex. Grossmann and Floudas [16] relaxed this assumption by developing a mixed-integer nonlinear programming (MINLP) problem for the feasibility test presented by Eq. 1.3. They further proposed an active constraint strategy for the solution of the resulting MINLP. The mixed-integer formulation also provides a systematic approach to consider all possible critical uncertainty realizations without exhaustive enumeration. The proposed formulation was utilized for synthesis of a heat exchanger network with uncertain stream flow rates and temperatures [17]. The case of linear constraints reduces to an MILP problem, for which global solution is attainable by standard branch and bound enumeration techniques [16, 18, 19]. Bansal et al. [20] developed a computationally efficient theory and algorithm based on multi-parametric programming techniques for this special case of flexibility analysis problems. The authors derived explicit expressions for the flexibility index as explicit functions of the continuous design variables. Pistikopoulos and Grossmann (1988a, 1988b, 1988c) used the flexibility test with linear constraints for optimal retrofit design [21–24] and redesign under infeasible nominal uncertainties [25]. Although these approaches are effective and promising to handle the design uncertainty, they require solving nested optimization problems, which poses a major challenge to solve complex and large scale problems in a reasonable time. Raspanti et al. [26] proposed replacing the complementarity conditions of the lower level optimization problems with a well-behaved, smoothed nonlinear equality constraints, namely Kreisselmeier and Steinhauser function [27] and Chen and Mangasarian smoothing function [28].

One of the common assumptions in these approaches is the known bounds of the uncertainties, which is rarely the case in real world industrial applications. Pistikopoulos and Mazzuchi [29] and Straub and Grossmann [30, 31] extended the flexibility test by assuming a probability distribution model for the parameter uncertainty, which improved the economical performance of the design optimization problem by addressing the “conservativeness” of the solution.

Another common assumption of these approaches is the steady state operation of the plant design, which creates a significant limitation on the applicability of the methodologies. Although steady state assumption holds true for the dominant life cycle of the plant operation, design optimization problem may fail to ensure the operability under transient behaviors such as startup or shutdown and transitions between different operating conditions. Dimitriadis and Pistikopoulos [32] proposed a dynamic feasibility index for the systems that are described by differential algebraic equations (DAE) subject to time-varying constraints. However, the time-dependent uncertainty in their formulation dictates to solve infinitely many dynamic optimization problems. Therefore, the authors assumed that the critical scenarios of uncertainties are known and lie on the vertices of the time-varying uncertainty space, similar to Swaney and Grossmann [14]. The simplifying assumption reduced the problem to the form given by Eq. 1.5.

$$
\begin{aligned}
DF(des) = \max_{\delta,u(t),t} \quad & \delta \\
s.t. \quad & \dot{x} = f(x(t), u(t), des, \theta(t), t), \quad x(0) = x_0 \\
& g(x(t), u(t), des, \theta(t), t) \leq 0 \\
& \theta(t) = \theta^N(t) + \delta\Delta\theta^c(t) \\
& \delta \geq 0, \quad \underline{u}(t) \leq u(t) \leq \overline{u}(t)
\end{aligned}
\tag{1.5}
$$

where the time dependence of the variables constitute a dynamic optimization problem, and the solution was determined by control vector parameterization techniques [32]. Dynamic flexibility has been widely utilized in numerous design optimization applications including batch processes [33], separation systems [34–38], reaction systems [39], and heat exchanger network synthesis [40–42].

The dynamic assessment of the plant feasibility under uncertainty has been also studied through exploiting the multiperiod design optimization formulation proposed by Halemane and Grossmann [13]. Varvarezos et al. [43] implemented an outer-approximation approach to solve the multiperiod multiproduct batch plant problems operating with single product campaigns, which was formulated as an MINLP. Pistikopoulos and Ierapetritou [44] considered stochastic process uncertainty and proposed a two-stage decomposition that can handle convex nonlinear problems.

As presented in this section, the early studies on integrated design optimization have primarily focused on (i) investigating the range of operation (flexibility) of a nominal design configuration under foreseeable conditions, and (ii) determining the "best" possible trade-off between the investment cost and the capability of handling variations in the internal and external operating conditions. These studies mostly considered open loop processes, under the traditional assumption that controller design is a sequential task to process design. However, most processes in industry are operated in closed loop, and the controller schemes inherently alter the process dynamics, rendering the open loop flexibility analyses of lesser relevance. In other words, an "attainable" operating point according to open loop flexibility analysis may actually be an infeasible point in closed loop. Realizing the shortcomings of open loop flexibility analyses, researchers began investigating the "controllability" of process systems, and the interdependence of process control and design decisions. In the following section, we present a retrospective background on the integration of process control in the design optimization problem.

1.3 Integration of Process Control in Design Optimization

The initial efforts towards the integration of process control and design problems established a fundamental understanding on the interdependence of the two decision making mechanisms. The most pronounced school of thought in the early years to evaluate the controllability of the process design is “dynamic resilience”, as conceptually defined by Morari [45, 46].

Morari [45] described dynamic resilience as “the ability of the plant to move fast and smoothly from one operating condition to another and to deal effectively with disturbances”. This depiction implies that there is not a clear cut distinction between flexibility, which was discussed in Sect. 1.2, and resilience. However, Grossmann and Morari [47] pointed out the main difference as “resiliency refers to the maintenance of satisfactory performance despite adverse conditions while flexibility is the ability to handle alternate (desirable) operating conditions”. This distinction is the primary motive for the majority of the flexibility analyses to study steady state operations, while the resilience deals with the dynamic operations, as we will discuss in this section.

Dynamic resilience, as described by Morari [45], aims to find the “perfect controller” that is allowed by the physical limitations of the system to assess the controllability of the process by using the internal model control (IMC) structure. The proposed technique decomposes the system transfer function $\tilde{G}$ into (i) a non-singular matrix $\tilde{G}_-$ to design the perfect controller $\tilde{G}_-^{-1}$, and (ii) a singular matrix $\tilde{G}_+$ to generate dynamic resilience indices based on (i) bounds on control variables, (ii) presence of right half plane transmission zeroes, (iii) presence of time delays, and (iv) plant-model mismatch. The proposed indices were utilized to improve the operability of numerous process, including heat integrated reactor networks [48–50], separation systems [51], heat exchanger networks [52].

Among the four aforementioned resilience indices, Perkins and Wong [53] studied the last two by adapting the “functional controllability” theorem proposed by Rosenbrock [54]. The authors further define a system to be functionally controllable if there exists a manipulated action $u(t)$ that can generate any process output $y(t)$ at any time t. Psarris and Floudas studied the dynamic resilience and functional controllability of multiple input multiple output (MIMO) closed loop systems with time delays [55–57], and transmission zeroes [56, 57]. Barton et al. [58] investigated the open loop process indicators, namely minimum singular value and right half plane zeros, to assess the interactions between different design configurations and their operability with the best possible control configurations.

In the context of simultaneously assessing the process controllability in process design, one of the first significant contributions is the “back-off approach” introduced by Narraway et al. [59]. Narraway and Perkins [60] used this approach to systematically assess the trade-offs between all possible controlled and manipulated variable pairs in a mixed integer formulation. Bahri et al. [61] employed the back-off approach to handle process uncertainties in an optimal control problem. The proposed approach is applicable to design linear and mildly nonlinear processes, and relies on three key steps, namely (i) perform a steady state nonlinear process optimization, (ii) linearize the process at the optimum point, and (iii) “back-off” from the optimal solution by some distance to ensure the feasibility of the operation under some structured disturbance profile. The proposed approach was shown to be effective effective to select between alternative flowsheets as well as alternative control structures.

With the burgeoning interest in exploring the simultaneous design and control problem, the International Federation of Automatic Control (IFAC) organized the first workshop on "Interactions between Process Design and Process Control" in the Centre for Process Systems Engineering at Imperial College London in 1992. The workshop laid the groundwork for a plethora of approches with a wide range of diversity. Walsh and Perkins [62] implemented a PI loop in the flexibility analysis, where the input-output loop is selected by an exhaustive screening procedure. Luyben and Floudas [63] formulated a multiobjective MINLP problem to simultaneously consider the disturbance rejection capacity of the control loop through disturbance condition number and relative gain array to evaluate the interactions between the inputs and outputs of a MIMO system, while designing the process. Shah et al. [64] used the State-Task Network (STN) representation [65] to simultaneously consider the scheduling and design problems in a batch plant. Thomaidis and Pistikopoulos [66] introduced a framework to consider the design problem simultaneously with (i) the process flexibility through stochastic flexibility index, (ii) the effect of equipment failures to the overall performance by combined flexibility-reliability index, and (iii) the impact of equipment availability by combined flexibility-reliability index. These aforementioned novel approaches were shown to be promising concepts and techniques to address multiple facets of operational decisions simultaneously with the process design problem. As a result, succeeding studies after this workshop expanded these techniques and branched out to explore further opportunities.

Integrating PI controllers in the design optimization problem was one of the prominent outcomes of the workshop and became the most attractive option for the following research. The literature on PI controllers was already abundant and well-established by the time. Moreover, the explicit form of the controller structure made the integration relatively easy and intuitive, which significantly accelerated the research in closed loop design optimization. Walsh and Perkins [67] presented an integrated PI control scheme and process design for waste water neutralization. Although the proposed approach was effective for the SISO process, it was reported that it entails further challenges for more complex processes. One major drawback of PI control is its inability to tackle MIMO systems without any advanced modifications in the feedback loop structure. Narraway and Perkins [60, 68] developed an MILP based formulation to systematically evaluate the economical performance of every input-output pair combination. Luyben and Floudas [69, 70] adapted a similar approach in a multiobjective framework to determine the best performing input-output pair based on the controllability indices introduced by them, earlier (1992) [63]. The proposed framework was showcased on the design of a heat integrated distillation system [69] and a reactor—separator—recycle system [70]. Mohideen et al. [34] formulated a multiperiod design and control problem to account for the dynamic variations in the operation, while including the input-output pairing superstructure in the problem. Moreover, the authors utilized the flexibility index to account for the uncertain parameters in the model and presented a decomposition algorithm for the resulting complex problem. Bansal et al. [71] constructed a similar formulation as a mixed-integer dynamic optimization (MIDO) problem, which was solved by a Generalized Benders Decomposition (GBD) based algorithm. The MIDO formulation was presented as follows.

$$
\begin{aligned}
\min_{u,des} \quad & \sum_{i \in NS} w_i C\big(\dot{x}^i(t), x^i(t), u^i(t), des^i\big) \\
s.t. \quad & \dot{x}^i(t) = h_d\big(x^i(t), u^i(t), des^i, \theta^i, t\big), x(t) = x_0 \\
& y^i(t) = h_a\big(x^i(t), u^i(t), des^i, \theta^i, t\big) \\
& g\big(\dot{x}^i(t), x^i(t), y^i(t), u^i(t), des^i, \theta^i, t\big) \leq 0
\end{aligned}
\tag{1.6}
$$

where w_i is the discrete probability of a scenario i and NS is the discretized set of scenarios. The discretization of uncertainty in the process was first proposed by Grossmann and Sargent [11].

Although the aforementioned PI based design and control frameworks are applicable on nonlinear processes, the range of operability is usually limited due to the mismatch between the nonlinear process model and the linearized control model. Ricardez Sandoval et al. [72, 73] used robust control tools and the back-off approach to integrate PI control and ensure its stability while solving the design optimization problem. The proposed approach was also tested against the Tennessee Eastman process [74]. The back-off approach was later generalized for control structure selection in nonlinear processes by Kookos and Perkins [75]. Ricardez-Sandoval & co-workers have extensively studied back-off approach for simultaneous process design and control under uncertainty [76–78].

One main limitation of integrating PI control in the design optimization in a dynamic formulation is the increasing problem size and complexity. Kookos and Perkins [79] developed an algorithm for the integrated PI control and design optimization problem, where the size of the search space was reduced systematically in each successive iteration. Malcolm et al. [80] proposed an "embedded control optimization" procedure, where the authors introduced a two-stage decomposition scheme that approximates the complete integrated problem. The proposed approach reduced the problem size and complexity, and was showcased on larger scale problems including a reactor—separator system [81].

Apart from the inability to naturally handle MIMO systems, PI controllers do not explicitly account for any process constraints stemming from operational, environmental, and safety limitations. Model predictive control (MPC) overcomes these shortcomings by postulating a constrained dynamic optimization problem subject to an explicit model of the process [8]. One of the first remarkable efforts to integrate an MPC scheme in a nonlinear design problem was published by Brengel and Seider [82]. Here, the authors postulate a bi-level optimization problem, where the *leader* has an economic objective, while the *follower* is the MPC formulation, as presented by Eq. 1.7.

$$
\begin{aligned}
\min_{des} \quad & C_{des}\big(des\big) + \kappa C_{\kappa}\big(x(t), y(t), u(t), des, \theta(t)\big) \\
s.t. \quad & f_{des}\big(des, \theta(t)\big) = 0 \\
& g_{des}\big(des, \theta(t)\big) \leq 0 \\
& \min_{u(t)} \quad C_u\big(x(t), y(t), u(t), des, \theta(t)\big) \\
& s.t. \quad \dot{x} = f_u\big(x(t), y(t), u(t), des, \theta(t)\big) \\
& \qquad g_u\big(x(t), y(t), u(t), des, \theta(t)\big) = 0 \\
& \qquad h_u\big(x(t), y(t), u(t), des, \theta(t)\big) \leq 0
\end{aligned}
\tag{1.7}
$$

where κ is the design and control integration parameter that scales the trade-off between the controllability of the system and the investment cost. The bi-level problem presented in Eq. 1.7 is challenging to solve without appealing to simplifications. Therefore, the authors proposed replacing the follower problem by complementary slackness equations. However, the solution strategy was still intractable for more complex systems due to the numerical calculation of the second derivatives [82]. As a consequence, integration of the MPC scheme in the design optimization had been rather limited in the literature for almost a decade, until the invention of multi-parametric MPC (mpMPC/explicit MPC).

Bemporad et al. [83] proposed formulating the MPC problem as an explicit function of the initial conditions of the system. This novel strategy allowed for deriving piecewise affine explicit control laws by treating the initial conditions as parameters. The proposed approach formulated the explicit MPC problem as presented by Eq. 1.8.

$$
\begin{aligned}
u_t(\theta) = \quad & \arg\min_{u_t} \|x_N\|_P^2 + \sum_{t=1}^{N-1} \|x_t\|_Q^2 + \sum_{t=1}^{N-1} \|y_t - y_t^{sp}\|_{QR}^2 + \sum_{t=0}^{M-1} \|u_t - u_t^{sp}\|_R^2 + \sum_{t=0}^{M-1} \|\Delta u_t\|_{R1}^2 \\
s.t. \quad & x_{t+1} = Ax_t + Bu_t + Cd_t, \quad y_t = Dx_t + Eu_t + Fd_t \\
& \underline{x}_t \leq x_t \leq \overline{x}_t, \quad \underline{y}_t \leq y_t \leq \overline{y}_t, \quad \underline{u}_t \leq u_t \leq \overline{u}_t, \quad \underline{\Delta u}_t \leq \Delta u_t \leq \overline{\Delta u}_t, \quad \underline{d}_t \leq d_t \leq \overline{d}_t \\
& \theta = [x_{t=0}, u_{t-1}, d_t, y_t^{sp}, u_t^{sp}]^T
\end{aligned}
\tag{1.8}
$$

where N is the prediction horizon, M is the output horizon, superscript sp denotes set point, Q, QR, R, and $R1$ are the corresponding weight matrices determined by tuning, P is calculated by discrete algebraic Riccati equation, and $\|\cdot\|_{\psi}$ denotes weighted vector norm with a weight matrix ψ. Different than conventional MPC, Eq. 1.8 formulates the optimal control problem exactly and completely offline as a function of the set of parameters θ. The solution of this problem can be determined by multi-parametric programming techniques, which express the solution space as a piecewise affine function, as presented by Eq. 1.9.

$$
\begin{aligned}
u_t(\theta) =& K_n\theta + r_n, \quad \forall \theta \in CR_n \\
CR_n :=& \{\theta \in \Theta \mid CR^A\theta \leq CR^b\}, \quad \forall n \in \{1, 2, ..., NC\}
\end{aligned}
\tag{1.9}
$$

where CR_n is refered as a critical region and it is the active polyhedral partition of the feasible parameter space, Θ is a closed and bounded set, and NC is the number of critical regions.

The control law given by Eq. 1.9 reduces the complexity of solving an online optimization problem to a simple look-up table algorithm (also known as point location problem) and function evaluation, all of which are affine operations. Hence, the complexity of implementing an MPC scheme is similar to that of a PI controller.

Sakizlis et al. [84] exploited the explicit nature of the mpMPC solution in the context of design and control integration. The authors formulated a bi-level mixed integer dynamic optimization problem similar to Eq. 1.7, where the leader accounted for the investment and operating costs in the objective function subject to the dynamic high fidelity model, and the follower MPC problem was substituted by the affine control law Eq. 1.9. The proposed formulation offered an elegant and systematic methodology to reduce the complexity of the bi-level Eq. 1.7 into a single level dynamic optimization problem. However, the solution strategy still required repetitive linearizations and solving a multi-parametric programming problem at every iteration, which can be restrictive for large scale complex problems. Diangelakis et al. [85] alleviated that limitation by deriving a "design dependent offline controller", which allowed for solving a single MIDO problem after integrating the control law in the high fidelity model. Eliminating the linearization step and formulating a single synergistic design and control problem also improved the economical performance of the resulting process compared to the approach proposed by Sakizlis et al. [84]. The proposed formulation was also showcased on a tank, a continuous stirred tank reactor, and a residential scale combined heat and power unit. The cost effectiveness of the MPC integrated optimal design was also reported to be superior than PI integrated approaches in the literature. Diangelakis and Pistikopoulos [86] reported that the mpMPC integrated optimal combined heat and power unit operated more fuel efficient in closed loop than PI integrated design. Similarly, Sanchez-Sanchez and Ricardez-Sandoval [87] showcased a system of CSTRs, where the MPC integrated framework reduced both the operating and the investment costs compared to the PI control integrated approach.

One common aspect of the studies on simultaneous design and control optimization is the assumption that the process will be operated around the same steady-state point throughout the entire life cycle of the plant. However, the external plant conditions, such as market conditions, may dictate a considerably wider operating region with multiple steady state points [1]. The increasing competition among the businesses impacts the volatility of the market, which creates rapid fluctuations in the energy and raw material prices as well as their availability. Moreover, the demand rate on the product is also subject to considerable variations during the plant operation. Therefore, it is clear that there exists a "best" operating strategy under the knowledge available to the operator, which necessitates the operability of the plant across a wider range. For example, high production rates may be less profitable during the night time because of increased energy prices and hence, operating the energy intensive processes during the day time may reduce the operating costs. This indicates that the operating level of a processing unit might vary drastically by the choice of the operator. However, the integrated design and control frameworks discussed in this section usually assume a single operating point around which a controllability and flexibility analysis is conducted. Consequently, these frameworks do not attempt to provide any means of guaranteeing the operability of the process at different regions. In the next section, we will discuss several approaches that account for multiple operating regions in a plant, and their scheduling during the operational optimization.

1.4 Classification of Integration Methodologies

Engell and Harjunkoski [88] provide insights into the initial efforts made to integrate scheduling and control, along with the various challenges encountered, including technical, business-related, and psychological aspects. The classification of integrated scheduling and control (iSC) problems can be broadly categorized into two approaches: *top-down bottom-up*.

1.4.1 Top-Down Integration

The top-down approach in integrated scheduling and control (iSC) involves incorporating control decisions into the scheduling problem by integrating the dynamic process model or a control scheme like Model Predictive Control (MPC) or Proportional-Integral (PI) controllers. When the dynamic model is included as additional constraints in the scheduling problem, it is referred to as integrated scheduling and dynamic optimization [89]. Solving this iSC problem yields an optimal schedule and a corresponding open-loop control policy for implementation. For continuous systems, time-slot-based scheduling formulations are commonly used, where each slot represents production and transition times [90]. Batch systems utilize state equipment network (SEN) and state-task network (STN) structures for integration [91]. As the scheduling level involves discrete decisions, the iSC problem is typically formulated as a Mixed-Integer Dynamic Optimization (MIDO) problem. Since chemical processes are often described by conservation equations, the dynamic model includes nonlinear terms. Consequently, the MIDO problem is transformed into a Mixed-Integer Nonlinear Programming (MINLP) problem after discretizing the dynamic equations. However, the presence of nonlinear terms can significantly increase the complexity of the integrated model, especially when nonconvexities arise. As a result, solving the MINLP problem becomes computationally demanding and challenging for large-scale applications, necessitating practical and efficient solution strategies [5]. To alleviate computational costs, initialization of the solution procedure close to the optimal or sub-optimal solution can be beneficial. However, obtaining accurate initial guess values is generally difficult. Various techniques have been proposed to decompose, initialize, and reformulate the integrated problem for both batch and continuous chemical and manufacturing systems. These include approaches such as Bender decomposition [92] and community detection for continuous systems and bi-level programming for batch systems [93]. In the case of multi-product continuous systems, integer variables can be avoided by using a switching system formulation.

To incorporate the closed-loop behavior of the system in the integrated scheduling and control (iSC) problem, the Model Predictive Control (MPC) formulation can be added as constraints in the scheduling problem, leading to bi-level optimization formulations [94]. This approach has also been employed in the integration of design and control. However, integrating the MPC formulation increases the complexity of the model, particularly when considering integer decisions in the formulations. Various strategies have been explored to reduce the size of the integrated model. Several techniques aim to reduce the complexity of the integrated model, including linear state space representations [95], latent variables [96], piece-wise affine (PWA) model approximations [97], linear metamodels [98], and time scale-bridging models [99]. However, in some cases, extensive plant/MPC simulations are necessary to identify accurate low-order representations of the integrated model. Additionally, some of these reduction techniques may encounter difficulties when dealing with discrete variables. Applications of these reduction techniques can be found in multi-product continuous systems such as continuously stirred tank reactors (CSTR) and multitask batch systems. An alternative strategy involves deriving the optimality conditions of the MPC and integrating them into the scheduling model [100, 101]. This formulation introduces a set of complementarity constraints, requiring proper reformulation to avoid numerical issues. This approach has been applied to multi-product CSTRs, as demonstrated in the work by [102].

1.4.2 Bottom-Up Integration

The bottom-up approach in integrated scheduling and control (iSC) involves incorporating scheduling decisions into the control formulation. One common strategy is to use an economic objective function within the Model Predictive Control (MPC) scheme, resulting in an economic MPC (EMPC) formulation [103]. While EMPC has potential and attractiveness, stability and feasibility considerations need to be revised as they may differ from traditional MPC [104]. EMPC does not necessarily operate the system at a specified steady state or target value; instead, it optimizes the process economics in a time-varying manner, leading to frequently changing set-points and scheduling decisions based on process economics and external perturbations. Additionally, EMPC may not account for discrete decisions such as sequencing relations. Nonetheless, EMPC is regarded as a promising bottom-up approach for efficient iSC in various applications, including chemical process networks [105], inventory management [106], and continuously stirred tank reactors (CSTRs) [107].

Rescheduling schemes have been proposed as extensions of iSC formulations to handle disturbances and unexpected events. Some studies propose online solutions to the iSC problem to correct deviations from desired state values. However, computational time constraints may make this approach impractical for large-scale models. Consequently, techniques to reduce model complexity, such as linear state space representations and multi-parametric programming, have been applied to develop iSC schemes suitable for efficient online implementation [108]. Frameworks involving a nested decision-making structure with two loops have also been proposed for continuous and batch systems [109–111]. In the outer loop, rescheduling is performed by solving an iSC problem (either full-space or reduced order) to determine optimal set-point values, which are then tracked using either MPC or explicit MPC in the inner loop.

A state space representation has been developed for scheduling, allowing the application of control theory concepts such as Model Predictive Control (MPC) and Economic MPC (EMPC) for rescheduling and online scheduling tasks [128, 129]. This unified state-space formulation for integrated scheduling and control (iSC) considers the closed-loop properties of the system [130]. The state space representation provides a natural description of the scheduling process as a dynamic system, offering more accurate scheduling models when used individually or integrated with planning and/or control. This approach has potential applications in multi-product continuous and batch systems, although model reduction techniques may be necessary for large-scale problems.

While the closed-loop implementation of iSC decisions can handle uncertainties at the control level in a reactive or preventive manner, uncertainties in the iSC formulation have received limited attention. Promising frameworks for handling uncertainties in the iSC model include two-stage stochastic programming, chance-constrained optimization, and robust optimization [131]. These approaches have been tested on multitask batch systems with uncertain kinetic parameters and initial conditions, as well as on a multi-product continuously stirred tank reactor (CSTR) with uncertain demand [132]. However, solving the resulting formulations can be challenging and may require decomposition techniques like Bender decomposition. Another approach for addressing uncertainties is the use of back-off terms to approximate their effects. These terms can be estimated through Monte Carlo simulations. An iterative procedure is employed to update the back-off terms until a convergence criterion is met. This approach has been applied to multi-product continuous systems, including process design and multi-product, multi-unit batch systems [109–111]. However, for large-scale problems, the computational cost can be significant due to the intensive calculations involved in Monte Carlo simulations.

In recent years, the integration of scheduling and control (iSC) strategies has been observed in various industrial applications. Bindlish [113] demonstrated the simultaneous solution of scheduling, real-time optimization, and control of power for industrial co-generation plants. They employed a first-principle, steady-state, non-linear model that was continuously tuned with plant data. This model was utilized for online optimization, followed by the implementation of scheduler results using a model-predictive controller. A nested decision-making framework was applied by [115]. to an Air Separation Unit. The framework consisted of an inner loop with an online model-predictive controller and an outer loop as a closed-loop simulation optimization problem. This approach allowed for the inte-

Table 1.1 Industrial applications of integrated scheduling and control [112]

Contribution	Description
[113]	Scheduling, optimization, and control of power for industrial co-generation plants
[99]	Development of frameworks for integrating scheduling and control operations in manufacturing plants
[114–116]	Integration of scheduling and model predictive control in the air separation unit operations
[117]	Integration of planning and scheduling in hydrogen supply networks
[118]	Integration of scheduling and closed-loop control online for MMA polymer manufacturing process
[109–111]	Scheduling and control for chemical batch processes
[119]	Integration of scheduling and control for reactive distillation systems and wastewater treatment systems
[120]	Scheduling and control framework for the integration of renewable energy sources for fuel cell systems
[121]	Scheduling and control of CO2 emission capture from coal-based power plant
[122, 123]	Robust online scheduling for cascaded hydro-power systems
[124]	Online scheduling for large HVAC systems
[125]	Scheduling and control methods for supply chain management
[126]	Integrated dynamic scheduling policies for polyol processes
[127]	Integrated scheduling and control for energy storage buildings

gration of scheduling and control in a coordinated manner. Baldea et al. [99] illustrated the iSC solution strategy for manufacturing plants using a scheduling-oriented model-predictive controller. They incorporated a scale-bridging model that provided an explicit description of the closed-loop process dynamics. This approach allowed for a comprehensive integration of scheduling, control, planning, and design aspects. Table 1.1 provides an extensive and comprehensive list of industrial applications where scheduling, control, planning, and design have been integrated. It showcases the diverse range of industries and processes in which iSC strategies have been implemented. Table 1.2 presents a comprehensive classification of the different approaches that have been utilized in investigating iSC over the past few years. This classification helps to categorize and understand the various methodologies and techniques employed in the field of integrated scheduling and control.

Table 1.2 Grand classification of methodologies [112]

Contribution	Approach	Methodology
[120, 144, 145, 148]	Bottom-up	Simultaneous/Decomposition MIDO/MINLP open loop dynamics
[125, 129, 130, 146, 149–152]	Bottom-up	Economic MPC/control theory in scheduling problems
[89, 119, 153, 154]	Top-down	Simultaneous/Decomposition MIDO formulation to MINLP
[7, 95, 97, 108, 115, 121, 155–160]	Top-down	Integration via mpMPC/NMPC/fast MPC
[96, 99, 161–165]	Bottom-up	Data-driven surrogate model and MINLP scheduling, Scale bridging control formulation in scheduling
[26, 131, 166–168]	Bottom-up	Flexibility, feasibility, scheduling under uncertainty
[109, 169]	Top-down	MIDO formulations with Stochastic back-off formulation for uncertainty
[165]	Top-down	Decomposition algorithms with surrogate model and MILP scheduling
[170, 171]	Top-down	Multistage mixed-integer optimal control formulation
[98]	Bottom-up	Multi set-point mpMPC scheduling layer with MILP control layer
[172, 173]	Top-down	Simultaneous/decomposition algorithms using control/dynamics-aware scheduling models
[128, 174–176]	Bottom-up	Advanced control and MIP scheduling schemes

1.4.3 Integration of Operability and Control

There has been a growing interest in coordinating control with safety considerations, recognizing that traditional approaches to process safety, such as process design modifications, may overlook important factors that impact process operational safety. These factors include the complex multi-variable interactions of process components and variables, limitations on the capacity of control actuators, and the influence of safety or relief system responses on the effectiveness of the process control system. To address these challenges, researchers have proposed various approaches. One such approach involves incorporating thresholds based on a state-based Safeness Index, as developed by [133], to serve as triggers for safety system activation. By integrating these thresholds into the design of safety systems, coordination between safety and control can be achieved. The same Safeness Index, with different thresholds, can also be utilized in the design of model-predictive controllers (MPC) to enhance coordination between safety and control designs [134, 135]. Control designs have been developed by [136–143] to handle safety-related issues, particularly in the context of faults. These control designs aim to detect and respond to faults in the system, ensuring safe and reliable operation. The integration of safety and control has also been explored through dynamic risk assessment, which plays a crucial role in providing real-time evaluation and mitigation of potential hazards. By dynamically assessing risks, safety measures can be promptly adjusted or implemented in response to changing process conditions, thereby enhancing overall safety performance and ensuring the protection of personnel, assets, and the environment. The integration of safety considerations with process control represents a significant advancement in ensuring safe and efficient operation. By considering both the inherent risks associated with the process and the effectiveness of control measures, this integrated approach enhances overall safety performance and helps mitigate potential hazards in real-time.

Most of these methodologies presented in the classification handle the linearities in a great way but there is still a struggle in handling non-linearities [144, 145], and uncertainties within the systems [146]. There have been considerable works on handling nonlinear systems via methods like model reduction, surrogate, and hybrid modeling. Further, uncertainties are gaining prevalence in recent academic research. Another major challenge that is being addressed in the current-day research is the scaling up of the integration methodologies to the industrial stage.

Although significant progress has been made in integrating design, control, scheduling, and operational functions such as safety and resilience, it is important to note that a universally accepted methodology or protocol for the integration of these aspects does not currently exist. Furthermore, there is currently no commercially available software or prototype system that fully supports such integration. To address this challenge, [147] proposed the PAROC framework as a step towards resolving this issue. The PAROC framework provides a systematic approach to integrating the different aspects of process operations. It aims to bridge the gap between design, control, scheduling, and other operational functions. While the PAROC framework is discussed in detail in Chap. 3, it is important to note that further research and development are still needed to establish a comprehensive and widely accepted methodology for the integration of design, control, scheduling, and operational functions. The absence of a commercially available software or prototype system underscores the ongoing efforts to address this challenge and highlights the need for continued advancements in this field.

1.5 Towards the Grand Unification of Process Design, Scheduling, and Control

Process design, scheduling, and control problems are traditionally constructed to address different objectives and they span widely different time scales. In a nutshell, the plant design problem dictates the capacity of processing and it usually comprises the most uncertainty due to its years long life-cycle. The scheduling problem addresses the allocation of the resources and time, as well as the operating level of processing units and their maintenance based on some economical criteria over days/months long horizons. Lastly, the control problem maintains the performance of the plant, while satisfying any physical limitations such as the environmental and safety constraints. The discrepancy in the objectives and time scales creates a challenging problem to systematically evaluate and determine the optimal trade-off between different decision makers.

Process scheduling is more critical in batch operations than continuous operations, as the former are inherently dynamically operated. Accordingly, the initial efforts focused primarily on the batch processes for the integration of the operational optimization and design problems. Birewar and Grossmann [177] formulated NLP models to incorporate the scheduling decisions in the batch sizing and timing problem in a multiproduct plant for unlimited intermediate storage and zero wait policies. Shah et al. [64] tackled a similar problem by using the STN representation. White et al. [178] investigated the switchability of continuous processes between different operating points through formulating an optimal control problem that accounts for the terminal criteria and path constraints within a range of design parameters. Bhatia and Biegler [179, 180] formulated a dynamic optimization problem, where an economic objective function was subject to a dynamic high fidelity model of the process described by differential algebraic system of equations. The authors proposed a solution strategy based on discretizing the process model by orthogonal collocation over finite elements, followed by solving the resulting NLP by using a standard solver. The proposed modeling and solution strategy was shown to be promising to satisfy the path constraints, which is a crucial benefit for dynamic systems. Terrazas-Moreno et al. [3] extended this integration approach to account for the binary decisions in the scheduling problem by formulating a MIDO. Similar to Bhatia and Biegler [179, 180], the authors first discretized the problem by orthogonal collocation, followed by solving the resulting MINLP.

The early studies that explore the interactions between the scheduling and process control decisions have a significant role in shaping today's approaches for the integrated design optimization problem. In their excellent review article, Baldea and Harjunkoski [181] classified these attempts to integrate the scheduling and control decisions as (i) "top-down approaches", where the process dynamics and control elements are incorporated in a scheduling skeleton, and (ii) "bottom-up approaches", where the process economics are implemented in the plant-wide control decisions.

In terms of characterizing the transitions between different products in a single operating unit, Mahadevan et al. [182] introduced a unique "top-down" perspective on the operational optimization problem, revealing that a simultaneous approach on the scheduling and control problem can identify and eliminate the fundamental limiting behavior during the transitions, as showcased on a polymer grade transition process. However, the presented approach requires case specific heuristic decisions to select the "best" fitting scheduling and control configuration and hence, it is not suitable for different applications in the general sense. Chatzidoukas et al. [183] studied a similar polymerization reactor, and formulated a MIDO problem to determine the time optimal transition between different polymer grades and best performing control structure simultaneously. Flores-Tlacuahuac and Grossmann [90] introduced a monolithic approach on a multiproduct cyclic CSTR, where the profit was maximized by manipulating the production sequence, transition times, production rates, length of processing times, and amounts manufactured of each product. Different from the earlier studies [182, 183], the authors focused on the manipulated actions rather than the optimal control configuration. They formulated a MIDO problem, which was solved by discretization of the differential algebraic equations by orthogonal collocation on finite elements followed by solving the resulting MINLP. The presented approach has been extensively studied in the following years to broaden its scope and effectiveness. Terrazas-Moreno et al. [184] applied this approach on two industrial polymerization reactors. Terrazas-Moreno et al. [3] formulated a design optimization problem accounting for the scheduling and open loop control trajectories using this approach. Flores-Tlacuahuac and Grossmann (2010, 2011) extended the formulation to partial differential equation systems, and showcased on tubular reactors with single [185] and multiple production lines [186].

This monolithic approach usually generates open loop control trajectories, i.e. no feedback loop is assumed to develop the input and output profiles. However, the processing units are subject to internal process disturbances, and the mismatch between the process and the model leads to deviations in the targeted operations. Zhuge and Ierapetritou [153] implemented the monolithic approach in closed loop, where the authors initiate a readjustment procedure to solve the integrated problem online if the states deviate from their reference trajectories. This approach does not completely resolve issue of handling the process disturbances or the process/model mismatch, however it was shown to mitigate these concerns to a great extent. Gutiérrez-Limón et al. [187] also implemented a similar closed loop strategy with a nonlinear model predictive control scheme, while extending the scope of the problem statement to account for an extended horizon production policy. However, both approaches require solving a complex and large scale MINLP problem at the time steps of the controller, which makes it unsuitable for the processes with fast dynamics.

Low-order representation of fast process dynamics in the scheduling problem has been an effective approach to reduce the computational burden of solving complex optimization problems. Du et al. [162] proposed a time scale-bridging model that describes the closed-loop inputâŁ"output behaviour of a process in the scheduling formulation, postulated as a MIDO problem. The low-order representation also maintains the stability of the process in the existence of process/model mismatch and handles disturbances. Baldea et al. [99] extended this approach to MPC governed systems.

Burnak et al. [95] also addressed the online computational burden of "top-down" approaches by developing a multi-parametric programming based approach, where the authors explicitly mapped (i) the closed loop dynamic process behavior in a "control-aware" scheduling problem, and (ii) the continuous and binary scheduling level decisions such as the operating level and operational mode of the system in a "schedule-aware" MPC scheme (iii) to yield the optimal operational decisions. The offline nature of the integrated scheduling and control scheme allows for determining the feasible operating space prior to actualizing the operation. Furthermore, reducing the problem complexity from solving online optimization problems to a simple look-up table and affine function evaluation, the framework is well-suited for fast process dynamics. Charitopoulos et al. [98] employed a similar multi-parametric programming approach to include the planning decisions in their framework.

In the "bottom-up" approaches, on the other hand, incorporating the economical objectives in the plant control structures has been perceived as the key for seamless integration of scheduling and control. For this purpose, MPC formulations provide the flexibility to account for a spectrum of objectives in the control level due to their optimization based structures. Loeblein and Perkins [188] presented an economical analysis of unconstrained MPC scheme operating under constrained systems. The authors determined the most cost effective model predictive regulatory control structure by utilizing the back-off approach to satisfy the constraints. Zanin et al. [189] addressed the discrepancy between the real-time optimization (RTO) and control layers by incorporating the economic optimization problem in the controller and feeding the same piece of information in both layers. The proposed formulation diminishes the discrepancy between the decision layers to yield more economical operations, but the resulting control scheme does not guarantee the stability of the process for the entirety of operations. Rawlings and Amrit [190] developed asymptotic stability criteria by formulating the so-called "economic MPC" (or EMPC), where the objective function of the MPC is designed to minimize the operational costs instead of maintaining the steady state of the process. This approach aims to replace the conventional two layer structure with RTO and dynamic regulatory control by a single control layer, where the economic optimization and process regulation are conducted simultaneously. Amrit et al. [191] further extended the stability criteria by (i) imposing a region constraint on the terminal state instead of a point constraint, and (ii) adding a penalty on the terminal state to the regulator cost.

Similar to the monolithic "top-down" scheduling and control approach, EMPC has been shown to be too complex to be solved in the control time steps. This limitation has led the researchers to develop decomposition algorithms for faster computational times. Würth et al. [192] proposed a decomposition framework for the single layer dynamic RTO formulation, where the slow trends and process uncertainty is handled in the upper layer, while the lower layer accounts for the fast disturbances actiong on the process. Ellis and Christofides [193] focused on selecting a suitable input configuration for such two-layered dynamic RTO structures such that the asymptotic stability is guaranteed. Jamaludin and Swartz [194] and Li and Swartz [195] employed a convex MPC problem in the lower level regulatory control, which enabled its exact substitution with KKT optimality conditions. Simkoff and Baldea [100] used the same substitution strategy on a production scheduling problem.

Design optimization accounting for the scheduling and control decisions with closed loop implementation is relatively recent in the literature. Patil et al. [4] modeled the product transitions in design optimization, while maintaining the stability of the closed loop system governed by a PI control scheme. The authors formulated an MINLP similar to Eq. 1.6 with the contribution of the criterion, $eig(A_i^z(x_{lin})) < 0$, which enforces the stability of the linearized states for all products i in a multiproduct unit under all critical scenarios z. Due to the linearization of the controllers around the operating point, this approach requires repetitive identification of the states at every optimization iteration. Koller and Ricardez-Sandoval [5] improved this approach by applying orthogonal collocation on finite elements on the integrated problem, and Koller et al. [6] employed the back-off method to satisfy the constraints under uncertainty by using Monte Carlo sampling techniques to determine the back-off terms.

Recently, Burnak et al. [7] introduced a multi-parametric programming based theory and framework for the integration of process design, scheduling, and control. The authors derived offline design dependent control and scheduling schemes that can be incorporated in a MIDO formulation in a multi-level fashion, as presented by Eq. 1.10.

$$
\begin{aligned}
\min_{u,s,des} \quad & \int_0^{\tau} C(x(t), y(t), u(t), s(t), des, d(t))dt \\
s.t. \quad & \dot{x}(t) = f(x(t), y(t), u(t), s(t), des, d(t), t) \\
& \underline{y} \le y(t) = g(x(t), y(t), u(t), s(t), des, d(t), t) \le \overline{y} \\
& \underline{x} \le x(t) \le \overline{x}, \quad \underline{des} \le des \le \overline{des}, \quad \underline{d} \le d(t) \le \overline{d} \\
& s_t(\theta_s) = \arg\min_{s} \quad \sum_{t_s \in N_s} C_s(x_{t_s}, y_{t_s}, s_{t_s}, des, d_{t_s}) \\
& \qquad\qquad s.t. \quad \underline{x}_{t_s} \le x_{t_s+1} = A_{t_s}x_{t_s} + B_{t_s}s_{t_s} + C_{t_s}d_{t_s} \le \overline{x}_{t_s} \\
& \qquad\qquad\qquad \underline{y}_{t_s} \le y_{t_s} = D_{t_s}x_{t_s} + E_{t_s}s_{t_s} + F_{t_s}d_{t_s} \le \overline{y}_{t_s} \\
& \qquad\qquad\qquad \underline{s}_{t_s} \le s_{t_s} \le \overline{s}_{t_s}, \quad \underline{d}_{t_s} \le d_{t_s} \le \overline{d}_{t_s} \\
& \qquad\qquad\qquad \underline{\theta}_s \le \theta_s = [x_{t_s=0}^T, y_{t_s=0}^T, d_{t_s}, des]^T \le \overline{\theta}_s \\
& u_t(\theta_c) = \arg\min_{c} \quad \sum_{t_c \in N_c} C_c(x_{t_c}, y_{t_c}, u_{t_c}, des, d_{t_c}) \\
& \qquad\qquad s.t. \quad \underline{x}_{t_c} \le x_{t_c+1} = A_{t_c}x_{t_c} + B_{t_c}u_{t_c} + C_{t_c}d_{t_c} \le \overline{x}_{t_c} \\
& \qquad\qquad\qquad \underline{y}_{t_c} \le y_{t_c} = D_{t_c}x_{t_c} + E_{t_c}u_{t_c} + F_{t_c}d_{t_c} \le \overline{y}_{t_c} \\
& \qquad\qquad\qquad \underline{u}_{t_c} \le u_{t_c} \le \overline{u}_{t_c}, \quad \underline{d}_{t_c} \le d_{t_c} \le \overline{d}_{t_c} \\
& \qquad\qquad\qquad \underline{\theta}_c \le \theta_c = [x_{t_c=0}^T, y_{t_c=0}^T, d_{t_c}, des]^T \le \overline{\theta}_c
\end{aligned}
\tag{1.10}
$$

where s and u denote the scheduling and control decisions, respectively. Note that the proposed formulation postulates explicit expressions for the scheduling and control strategies as functions of a set of parameters, θ, which includes the design of the process. The design dependence of the operational strategies allows for their direct integration in the MIDO formulation. The postulated formulation has two main benefits, (i) due to the explicit form of the follower problems, the multi-level MIDO problem is reduced to a single level, and (ii) only the design variables are left as the degrees of freedom of the problem, since the remaining are determined as a function of the design. This approach will be further explored in detail throughout this book.

1.6 Current Challenges and Future Directions

The PSE community has achieved unequivocally remarkable progress in realizing and advancing the set goals of Professor Sargent on systematic design optimization in five decades. Today, using design optimization tools to at least some extent has long become the standard practice in many industries. Commercial modeling and simulation software tools such as gPROMS[1] and Aspen Plus Dynamics[2] have been featuring robust and efficient solvers for dynamic optimization problems for a few years. Despite these milestones in PSE, we still have to make significant assumptions and simplifications regarding the operational decisions in the process design phase, even though the impact of their interdependence on process economics and operability has been articulated in numerous studies. Hence, the academia still needs to mature the theoretical foundations and the applicability of unified design optimization approaches before it gains wide industrial recognition. Here, we discuss some of the bottlenecks and potential directions to improve the state-of-the-art for industrial practice.

1.6.1 Need for an Industrial Benchmark Problem

As we have presented in this chapter, there is a plethora of proposed modeling techniques and solution approaches for the next generation unified design optimization problems. Therefore, it is clear that we need a generally accepted benchmark problem, preferably in industrial scale, to validate the effectiveness of proposed methodologies. The PSE community has benefited greatly from such standardized problems, such as the famous Tennessee Eastman Process detailed by Downs and Vogel [196] for process control studies. We believe that a well-defined problem will clarify the objectives in unified design optimization and accelerate the research towards industrial expectations. The problem should describe at least the following.

[1] https://www.psenterprise.com/products/gproms.

[2] https://www.aspentech.com/en/products/pages/aspen-plus-dynamics.

i. *A high fidelity model that describes the dynamics of the process.* The model should feature appropriate design variables to exhibit the dynamic consequences of scaling up/down the process. Furthermore, considering the reduction in capital investment that the multipurpose and multiproduct operating units provide, the process should comprise such units to examine the scheduling/design and scheduling/control trade-offs. Recent research that consider process design, scheduling, and closed loop control problems simultaneously [4, 6, 7] have studied only a single processing unit, which reflects a limited fraction of the overall benefit that the grand unification can provide.
ii. *Cost relations for investment, utility, and raw materials.* A functional form of the investment cost with respect to the capacity of the process is required to have standardized comparable results. Also, utility costs and raw materials may vary significantly, which inevitably impacts the optimal scheduling decisions. For instance, grid electricity costs are known to exhibit considerable differences during the day and night times. Thus, operational loads in energy intensive processes may fluctuate heavily. The impact of such changes in operating levels on design and control decisions were discussed in Sect. 1.3.
iii. *Product demand and availability of the utility, raw materials, and operating units over a time horizon.* Production allocation and timing is a key aspect of scheduling problem, which are heavily dictated by the product demand and availability of resources. However, it is not a trivial practice to estimate the future of these quantities. Therefore, probability distributions of these components will be beneficial to determine their expected values, while being able to take into account their worst case scenarios.

1.6.2 Robust Advanced Control and Scheduling Strategies

Incorporation of advanced control schemes seamlessly in the design optimization problem requires the controller to capture the dynamics of the process for the entire range of design variables. Burnak et al. [7] attempted to approximately model the design configuration as a right hand uncertainty in the constraint set, validated by closed loop simulations and closed loop MIDO problems. However, the design variables impose uncertainty in the left hand side of the constraints, as well as the nonlinear and bilinear terms in the objective function. Therefore, robust control strategies need to be developed for accurate predictions of future states in the control level prior to the realization of the design, and to guarantee the stability of the closed loop operations in simultaneous approaches.

Analogously, scheduling schemes should be robustified in the design optimization to minimize the rescheduling due to unexpected disruptive events, such as unit failure, drastic changes in product demand rate and raw material availability. Excluding these events in the scheduling scheme may result in steep changes in the target operation, and thus unattainable set points for the controller.

1.6.3 Flowsheet Optimization, Process Intensification, and Modular Design Opportunities

Optimization based plant design techniques have been used and developed for more than four decades [197, 198]. These techniques postulate "superstructures" that systematically simulate and compare every combination of flowsheet possibilities to determine the optimal process. More recently, superstructures have been formulated at the phenomena level to capture the fundamental relations between the mass and energy, which in turn yields intensified processes [199–207]. Such intensified processes are expected to deliver significantly increased operational efficiency and decreased unit volumes, making them very attractive options both in academia and industry [208]. This rapidly growing interest in intensified processes is one of the most pronounced directions that the PSE community has been taking. Therefore, studying these intensified processes in the context of unified design optimization will attract a wider audience from the industry. Clearly, modeling the spatial (synthesis/intensification) and temporal (scheduling/control) decisions simultaneously in a single problem formulation will capture even more synergistic interactions, which will increase the process profitability.

Furthermore, the researchers studying process intensification can benefit from the tools and methodologies on unification of design, scheduling, and control. Baldea [209] reported a theoretical justification for the loss of control degrees of freedom due to process intensification, which poses a significant limitation on intensification activities. Tian and Pistikopoulos [210] and Dias and Ierapetritou [211] discuss the limitations on the operability of such intensified systems and potential directions to overcome these limitations in their excellent review papers. The researchers on process intensification technologies can adopt the techniques, ranging from steady state and dynamic flexibility to integration of scheduling and control decisions, in order to address the operability issues.

1.6.4 Integration of Scales

The urgent need to reduce carbon emissions, minimize waste generation, and develop environmentally friendly processes has further prompted a shift towards more responsible and sustainable operations. By integrating quantitative methods, advanced modeling techniques, and sustainability metrics, frameworks, and tools that enable comprehensive decision-making, accounting for economic, environmental, and social dimensions are emerging [212]. The need for an integrated, holistic approach has never been more important, with the view to systematically coordinate the different layers of decision-making, analyze their interactions, and provide solutions that seamlessly look at the different scales of decision-making regarding control, scheduling, and ultimately design.

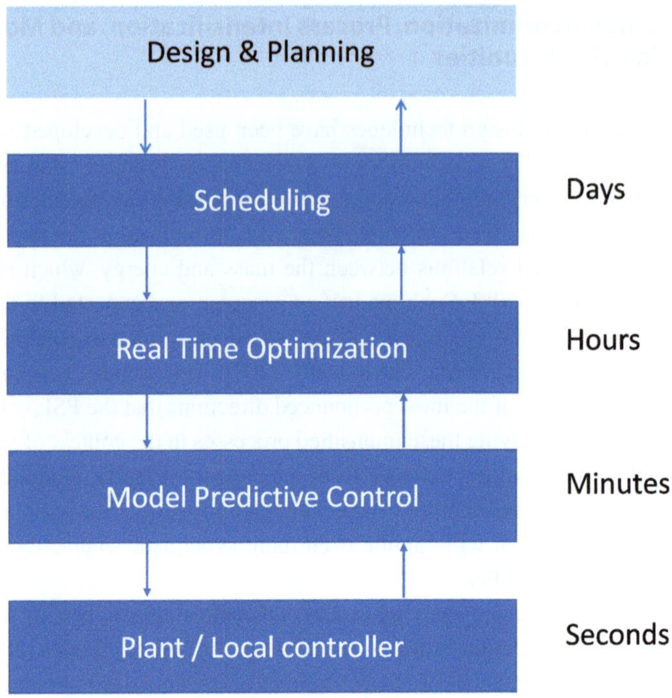

Fig. 1.3 The Conventional Decision-making Hierarchy offers a structured approach to optimize industrial processes. While it provides a logical sequence for decision-making, it also has limitations regarding sub-optimality and iterative adjustments

Traditionally, such operational decisions have been implemented over different time horizons ranging respectively from several days/weeks to minutes/seconds in a hierarchical manner, as shown in Fig. 1.3.

These decision layers are typically isolated and their interactions may be neglected. Each layer's objective(s), time scales, and sources of uncertainty are usually treated via different models and based on various data streams. On the other hand, the notion of Enterprise-wide optimization (EWO) focuses on harmonizing and optimizing operations at all levels, by explicitly recognizing the interdependencies and opportunities for the integration of information and decision-making across the diverse functions within a company's supply chain [213, 214].

Such integration of scales is typically driven by various factors such as smart manufacturing [215], real-time optimization [181], digital transformation [216], standardization, and increasing need for safety, sustainability, and cost-competitive resilience [212]. While such integration of scheduling, control, and design may facilitate improved decision-making and operational performance, it presents numerous challenges attributed to various factors, including:

- *Interdisciplinary Collaboration:* Integrating scheduling, design, and control requires collaboration among experts from different disciplines, including operations research, control engineering, and system design. Bridging the gaps between these disciplines and establishing effective communication and understanding can be challenging.
- *Complexity and Scale:* Real-world systems are often complex and large-scale, involving numerous components, constraints, and interactions. Managing the complexity and developing scalable integration approaches that can handle the sheer volume of data and decision variables is a significant challenge.
- *Dynamic and Uncertain Environments:* Many systems operate in dynamic and uncertain environments where parameters, demands, and constraints change over time. Developing integration methods that can adapt and respond to such dynamic changes, while considering uncertainty, is a complex task.
- *Conflicting Objectives:* Scheduling, design, and control often involve optimizing multiple conflicting objectives. For example, scheduling decisions may aim to minimize the makespan, while control decisions focus on stability and efficiency. Balancing these conflicting objectives and finding optimal or near-optimal solutions is a challenge.
- *Real-Time Decision-Making:* Integrating scheduling, design, and control in real-time decision-making scenarios requires efficient algorithms and methods that can process large amounts of data and provide timely solutions. Achieving low computational complexity and real-time responsiveness is crucial.
- *Information and Data Integration:* Integration requires the seamless exchange and integration of information and data across scheduling, design, and control domains. Ensuring the compatibility, consistency, and accuracy of data from different sources and formats is a challenge.
- *Software and System Integration:* Integrating scheduling, design, and control often involves the integration of different software tools and systems. Ensuring compatibility, interoperability, and efficient communication between these systems can be challenging, especially when they are developed by different vendors or based on different technologies.
- *Human Factors:* Integrating scheduling, design, and control systems also involves considering human factors, such as user interfaces, decision support systems, and human-machine interactions. Designing intuitive and user-friendly interfaces that facilitate effective human involvement in decision-making is a challenge.
- *Implementation and Deployment:* Transitioning from research and development to practical implementation and deployment of integrated systems is a challenge. Incorporating the integrated approach into existing systems, managing system upgrades, and ensuring smooth operation require careful planning and coordination.
- *Cost and Resource Constraints:* Integrating scheduling, design, and control should consider practical constraints, such as limited resources, budget constraints, and cost considerations. Finding cost-effective solutions that optimize system performance while respecting these constraints is a challenge.

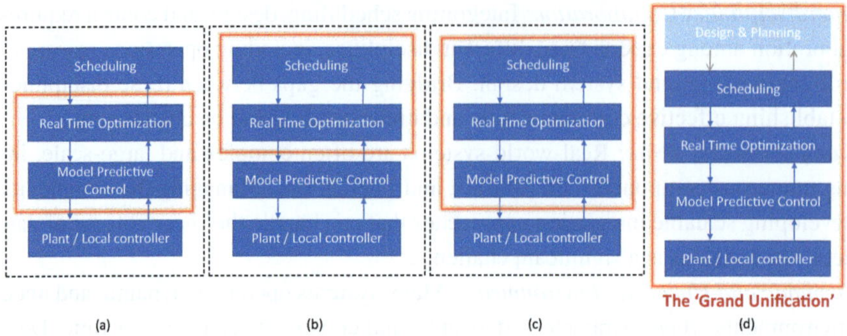

Fig. 1.4 Various opportunities for integration of scales, which facilitates a comprehensive understanding of the system and enables better alignment of decisions with strategic goals, operational requirements, and broader socio-economic and environmental considerations. **a** Integrating a real-time optimization layer that considers real-time measurements to dynamically adjust control variables and optimize system performance within predefined constraints and model predictive control layer thereby having hours and minutes time scale into consideration **b** Integrating scheduling layer which operates in days with real-time optimization layer which operates in hours **c** Integrating the top three layers of the conventional hierarchy thereby allowing the impact of lower level changes on the higher level and vice versa **d** The Grand unification of layers which poses the ultimate complexity in integrating and also expects to provide the best decision-making solutions

Addressing these challenges requires a comprehensive and systematic approach that considers the specific requirements and characteristics of the integrated system. This alongside the increasing need for the integration of scales led to various attempts in the development of generalized step-by-step frameworks. There are various methodologies proposed for such an integration, as discussed in this book (Fig. 1.4).

1.6.4.1 Integration of Scheduling and Control

Scheduling is a critical decision-making process with significant importance in process industries. Its primary objective is to efficiently allocate limited resources to processing tasks over time, ensuring the fulfillment of demands and facilitating profitable operations. These scheduling problems can be modeled based on the specific characteristics and constraints of the process, allowing for the proposal of different objective functions such as maximizing profit or minimizing make-span. Through this modeling, the optimal sequence of tasks and resource allocation can be determined.

On the other hand, process control also serves as a decision-making process aimed at managing the qualitative aspects of processes by adjusting their operating conditions. Advanced control techniques are gaining prominence across industries and are recognized by major companies as key elements for energy conservation, maximizing throughput, and achieving cost-efficient production [166, 181].

Traditionally, production scheduling and dynamic optimization problems in chemical processes have been approached separately. However, there exist variables that establish a connection between these two areas. For instance, the states at the beginning and end of transient periods, the duration of transient and steady-state periods, and the quantity of material in production stages are all linking variables. This linkage between scheduling and dynamic optimization implies that they cannot be completely isolated from each other [217]. Another reason why integrating scheduling and control problems are challenging is due to the fundamental difference in the time horizons they address. Scheduling focuses on long-term calculations that span hours or days, while control actions operate on much shorter time scales, usually within seconds to a few minutes. This disparity in time frames presents a significant obstacle when attempting to combine scheduling and control processes [181]. By adopting an integrated approach, information can be shared between the scheduling and dynamic optimization levels, enabling integrated decision-making that enhances the profitability of process operations.

The integration of process scheduling and control poses several challenges. These challenges encompass the need for synchronization and coordination between scheduling and control decisions, the incorporation of real-time production data for dynamic decision-making, the complexity arising from resource constraints and uncertainties, and the integration of feedback control strategies into scheduling algorithms. Additionally, challenges related to system scalability, computational efficiency, and the integration of advanced technologies such as artificial intelligence and machine learning are also explored.

Current research emphasizes the considerable advantages associated with integrating process scheduling and control within manufacturing environments. The integration of these systems enables the reduction of production lead times by optimizing the sequencing and coordination of operations. By incorporating real-time data and feedback control, it becomes possible to enhance resource utilization, improve product quality, and enhance the adaptability of production processes to changing circumstances. Consequently, this integration contributes to improved manufacturing efficiency, increased throughput, and heightened customer satisfaction. Extensive ongoing research in this field, along with a focus on process design, is evident and can be observed in Table 1.1 and 1.2.

In the pursuit of effectively integrating process scheduling and control, researchers have explored a wide array of methodologies and technologies. One line of investigation centers around the development of optimization models and algorithms that take a holistic approach to consider both scheduling and control objectives. By enabling synchronized decision-making, these models and algorithms aim to optimize the overall system performance by carefully examining the interplay between scheduling and control variables. Advanced control strategies, including model predictive control and adaptive control, undergo rigorous scrutiny to evaluate their effectiveness in managing dynamic production conditions and enhancing process performance. Leveraging real-time data and adaptive algorithms, these strategies dynamically adapt control actions and optimize process operations in response to changing circumstances.

Moreover, the research delves into the significance of real-time data acquisition and analysis, Internet of Things (IoT) technologies, and digital twins in supporting the seamless integration of scheduling and control systems [102]. Real-time data acquisition and analysis play a pivotal role in providing valuable insights into the present state of the production process, empowering timely decision-making. Through the facilitation of seamless communication and data exchange between various components of the manufacturing system, IoT technologies enable efficient coordination between scheduling and control functions. Additionally, digital twins, which serve as virtual replicas of physical assets and processes, prove to be invaluable tools in simulating and optimizing production operations. Leveraging digital twins, researchers and practitioners can conduct virtual experiments and optimize system performance within a risk-free environment, thus furthering the integration of scheduling and control. Further discussion on the aspect is provided in Chap. 5 and in the next sections.

1.6.4.2 Integration of Operabilty and Control

Process operability plays a crucial role in the efficient and successful operation of industrial processes. It encompasses a broad spectrum of factors that impact the ability to control, monitor, and maintain a process in a safe and reliable manner. Achieving effective process operability necessitates the integration of various elements such as design considerations, instrumentation, control systems, maintenance strategies, and human factors. By prioritizing operability, organizations can optimize production levels, mitigate risks, and ensure the well-being of personnel and the environment. Ongoing research in process operability focuses on developing and refining methodologies, technologies, and best practices that enhance the operability of complex systems. This includes advancements in process control algorithms, intelligent monitoring systems, risk assessment, and management techniques, as well as training programs and the design of human-machine interfaces. The aim is to continually improve the operability of processes, ensuring their stability and reliability under various operational conditions. Table 1.2 provides a comprehensive list of academic research groups actively engaged in studying the integration of control with operability. These groups contribute to advancing knowledge and expertise in the field, driving innovation, and promoting the adoption of effective operability practices across industries.

Traditionally, process control is responsible for regulating and optimizing process variables, while process operability ensures efficient and safe operation. However, recent research has shed light on the potential advantages of integrating these two aspects to achieve superior industrial performance. Current research studies explore the synergistic relationship between process control and operability, examining the potential benefits that arise from their integration [218]. The integration of process control and operability yields significant benefits. One key observation is that improved process stability leads to enhanced product quality, increased production efficiency, and reduced variability. By integrating these two aspects, organizations can implement proactive maintenance strategies, mitigating the risk of equipment failures and unplanned shutdowns. Additionally, incorporating human

Table 1.3 An indicative list of operability and safety in literature [112]

Author (year)	Contribution
Pistikopoulos and co- (2018, 2021) [220, 221]	Challenges and opportunities in the integration of design, operability, and control of intensified and modular process systems, Integration of process intensification systems with guaranteed flexibility and safety performances
Lima and co- (2021) [222]	Development of novel control strategy using process operability and biometric control algorithms for batch bio-processes
Fischer and Douglas (1985) [223]	Evaluation of process operability during the initial design phase
Wang et al. (1996) [224]	Safety and operability analyses of processes using neural network learning algorithms
Georgiadis and -co (2002) [151]	Interactions of design, control, and operability in reactive distillation systems
Longwell (1994) [225]	Dynamic modelling for control and operability
El-Halwagi and co- (2006) [226]	Integration of inherently safer process design and scheduling
Christofides and co- (2018, 2019) [134, 135]	Safety via model predictive control

factors considerations into control system design enhances operator situational awareness, minimizing the likelihood of human error and improving overall safety [219]. By merging process control and operability, industrial processes can achieve heightened performance and optimize their operations. This integration paves the way for enhanced product quality, improved efficiency, reduced variability, proactive maintenance, and enhanced safety.

There are several challenges identified in integrating process control and operability. These include the need for compatible control and operability objectives, the availability and reliability of real-time process data, the development of robust control strategies that account for process disturbances, and the consideration of human factors in control system design. Challenges related to system complexity, scalability, and the integration of advanced technologies, such as artificial intelligence and machine learning, are also explored. The current-day research investigates various methodologies and technologies that facilitate this integration of process control and operability. This includes the development of advanced control algorithms, such as model predictive control and adaptive control, that consider operability constraints. Real-time monitoring techniques, such as fault detection and diagnosis systems, are examined for their ability to enhance process reliability and safety. The role of data analytics and predictive maintenance practices in improving operability and reducing downtime is also explored (Table 1.3).

1.6.5 Theoretical and Algorithmic Developments in MIDO

The most limiting bottleneck of the simultaneous approaches is the size of the integrated MIDO problems. The time component of the problem significantly increases the computational complexity, yielding infinitely many NP-hard problems to acquire an optimal solution profile. However, tailored algorithms can be developed by utilizing the special structure of such integrated problems. For instance, the open loop design optimization problem is relatively simpler than the integrated MIDO, and constitutes a lower bound on the optimal solution of the overall problem. Such properties can be exploited in decomposing the MIDO into subproblems to significantly reduce the search space for faster algorithms.

1.6.6 Software Development

Despite the theoretical and practical advances in the unified design problem among the academia, there is no commercially available platform or a software prototype. Such a tool will make the integrated approaches more accessible to the process designers in industry who are not necessarily experts on process control and scheduling, and it will attract more researchers from different disciplines and backgrounds. Pistikopoulos et al. [227] introduced the PARametric Optimization and Control (PAROC) framework to design explicit controllers based on high fidelity models, which can be a viable option to address the grand unification challenge [7, 85, 95, 147, 175]. However, it is clear that more progress is needed to engage a wider audience.

1.6.7 First-Principle Modeling in the Age of Artificial Intelligence

The release of ChatGPT in 2022 by OpenAI has brought significant attention to the power, scalability, and practical benefits of data-driven models, reinforcing their potential across various domains. The PSE community has long recognized these trends, as seen in tools like OMLT, which seamlessly integrates machine learning models into mathematical optimization frameworks. Similarly, conventional operations research (OR) tools, such as the Gurobi Optimizer, have embraced this shift by introducing Gurobi ML, enabling direct optimization over decision trees and neural networks. These advancements highlight the growing role of surrogate modeling and hybrid AI-optimization approaches, where data-driven models enhance computational efficiency and decision-making in complex systems.

2 Mixed-Integer Dynamic Optimization for Simultaneous Process Design and Control

Mixed-integer dynamic optimization (MIDO) problems arise in chemical engineering whenever discrete and continuous decisions are to be made for a system described by a transient model. Areas of application include integrated design and control, synthesis of reactor networks, reduction of kinetic mechanisms and optimization of hybrid systems. This chapter presents new formulations and algorithms for solving MIDO problems. The algorithms are based on decomposition into primal, dynamic optimization and master, mixed-integer linear programming sub-problems. They do not depend on the use of a particular primal dynamic optimization method and they do not require the solution of an intermediate adjoint problem for constructing the master problem, even when the integer variables appear explicitly in the differential-algebraic equation system. The practical potential of the algorithms is demonstrated on a distillation design and control optimization example.[1]

2.1 Overview

The last decades have seen a rapid increase in the use of dynamic simulation as a tool for studying the transient behaviour of process systems. In fact, common practice is to use dynamic simulation as a means towards "optimizing" process design and/or operation. This is achieved by an engineer performing multiple, trial-and-error simulations in order to choose a set of values for the process parameters that leads to "good" performance (e.g. low cost or fast transition between steady-states) while satisfying the constraints under which the process operates. Although this approach is simple and intuitive, it has the major drawback that the fraction of the parameter space that can be searched is severely limited, especially as

[1] Reprinted from Computers and Chemical Engineering, vol. 27, no. 5, V. Bansal, V. Sakizlis, R. Ross, J. D. Perkins, and E. N. Pistikopoulos, New algorithms for mixed-integer dynamic optimization, pp. 647–668, 2003, with permission from Elsevier [228].

B. Burnak et al., *Integrated Process Design and Operational Optimization via Multi-parametric Programming*, Synthesis Lectures on Engineering, Science, and Technology, https://doi.org/10.1007/978-3-032-10181-5_2

the number of parameters increases. This not only means that it is impossible to guarantee optimality, but in many cases it can even be difficult to find a set of parameter values that leads to a feasible solution where all the constraints are satisfied [229].

The ad hoc approach described above can be avoided by posing formal, dynamic optimization problems, where the objective is to select the set of time dependent and/or time-invariant parameters that optimize some aspect of the transient behaviour of the process while satisfying a given set of constraints. A number of algorithms have been developed in the literature for solving such problems and their reliability has evolved to the extent that realistic engineering problems involving thousands of variables can now be readily solved with commercial codes such as gPROMS.[2] Reported applications using gPROMS include optimization of operating policies in multi-component batch distillation and other processes [230–232], operation of catalytic reactors [233], design and operation of heat exchangers under fouling [234], analysis of hazards in batch processes [235], and the simultaneous optimization of the process and control system design in complex distillation systems [236]. Other notable applications of dynamic optimization to chemical engineering processes include online model predictive control [237, 238], fault diagnosis [239], determination of policies for switching between steady-states [178, 240, 241], and optimization of systems subject to exceptional events such as failures and sudden changes in product demand [242].

Practical engineering problems generally involve discrete as well as continuous decisions. For example, in the optimal design of a distillation column, the number of trays and feed location take discrete values, while the values of the column diameter and surface areas of the reboiler and condenser are continuous. This gives rise to mixed-integer dynamic optimization (MIDO) problems, the solution of which is a formidable task. The development of algorithms for solving such problems is still in its infancy and thus few large-scale applications have been reported. The objective of this chapter is to introduce new algorithms for solving MIDO problems that do not depend on particular integration and dynamic optimization solution methods, require significantly less intermediate information than current algorithms, and as such, are better suited to the solution of larger, more realistic problems.

The remainder of the chapter is structured as follows. After some mathematical preliminaries, the existing solution approaches for MIDO are thoroughly and critically reviewed. An alternative formulation and algorithm based on Generalized Benders Decomposition (GBD) principles is then presented and its advantages are discussed. An analogous algorithm based on outer-approximation (OA) is also outlined and, finally, the application of the two algorithms is illustrated with a distillation design and control example.

[2] https://www.psenterprise.com/products/gproms.

2.2 Preliminaries

A MIDO problem can be formulated as:

$$\begin{aligned}
&\min_{\mathbf{u}(t),\mathbf{d},\boldsymbol{\omega}} \; J(\dot{\mathbf{x}}_d(t), \mathbf{x}_d(t), \mathbf{x}_a(t), \mathbf{u}(t), \mathbf{d}, \boldsymbol{\omega}, t) \\
&s.t. \;\; \mathbf{h}_d(\dot{\mathbf{x}}_d(t), \mathbf{x}_d(t), \mathbf{x}_a(t), \mathbf{u}(t), \mathbf{d}, \boldsymbol{\omega}, t) = 0, \quad \forall t \in [t_0, t_f] \\
&\qquad \mathbf{h}_a(\mathbf{x}_d(t), \mathbf{x}_a(t), \mathbf{u}(t), \mathbf{d}, \boldsymbol{\omega}, t) = 0, \quad \forall t \in [t_0, t_f] \\
&\qquad \mathbf{h}_0(\dot{\mathbf{x}}_d(t_0), \mathbf{x}_d(t_0), \mathbf{x}_a(t_0), \mathbf{u}(t_0), \mathbf{d}, \boldsymbol{\omega}, t_0) = 0 \\
&\qquad \mathbf{h}_p(\dot{\mathbf{x}}_d(t_i), \mathbf{x}_d(t_i), \mathbf{x}_a(t_i), \mathbf{u}(t_i), \mathbf{d}, \boldsymbol{\omega}, t_i) = 0, \quad \forall t_i \in [t_0, t_f], i = 1, ..., N \\
&\qquad \mathbf{g}_p(\dot{\mathbf{x}}_d(t_i), \mathbf{x}_d(t_i), \mathbf{x}_a(t_i), \mathbf{u}(t_i), \mathbf{d}, \boldsymbol{\omega}, t_i) le0, \quad \forall t_i \in [t_0, t_f], i = 1, ..., N \\
&\qquad \mathbf{h}_q(\mathbf{d}, \boldsymbol{\omega}) = 0, \\
&\qquad \mathbf{g}_q(\mathbf{d}, \boldsymbol{\omega}) \leq 0
\end{aligned} \tag{2.1}$$

where $\mathbf{h}_d = 0$ and $\mathbf{h}_a = 0$ represent the system of differential algebraic equations (DAE) with initial conditions $\mathbf{h}_0 = 0$, $\mathbf{h}_p = 0$, and $\mathbf{g}_p \leq 0$ represent the set of point equalities and inequalities, respectively, that must be satisfied at specific time instances (including the end-point if necessary), $\mathbf{h}_q = 0$ and $\mathbf{g}_p \leq 0$ are the time-invariant equality constraints, $\mathbf{x}_d(t)$ and $\mathbf{x}_a(t)$ are the vectors of differential state and algebraic variables, d is the vector of time-invariant continuous variables, and $\boldsymbol{\omega}$ is the vector of time-invariant integer variables. In the sequel, we will consider $\boldsymbol{\omega}$ to be a vector of binary variables since most process engineering applications are modelled in this way and any integer variable can be expressed as a sum of binary variables [243]. The case of time-varying binary variables will be considered later in this chapter.

Note that the formulation in Eq. 2.1 is quite general. For example, integral terms in the objective function can be eliminated by the addition of extra differential equations. If there are path inequality constraints that must be satisfied at all times during the time horizon, these can always be converted into end-point inequality constraints [244] (e.g. Bansal, 2000, Sullivan and Sargent, 1979), which are a sub-set of $\mathbf{g}_p \leq 0$.

2.3 Review of Existing Solution Approaches for MIDO

By definition, it is the presence of the binary variables $\boldsymbol{\omega}$ that complicate the solution of a MIDO problem compared to a continuous dynamic optimization problem (for which solution techniques are well-established). In the context of optimally designing dynamic systems for robustness, Samsatli et al. [245] proposed the use of smooth approximations for the binary variables in order to convert the discrete-continuous problem Eq. 2.1 into a purely continuous one. For example, if a binary variable $\boldsymbol{\omega}$ is defined such that it takes a value of

1 if another (continuous) variable $x > 0$ and is zero otherwise, then the following smooth approximation would be used.

$$\omega = \frac{1}{2}\big(\tanh(\beta x) + 1\big) \tag{2.2}$$

where β is a large, positive number. The major drawback of trying to solve general MIDO problems using such an approach is that the smooth approximation does not always lead to integral values of $\boldsymbol{\omega}$ (e.g., in Eq. 2.2, when $x = 0$, $\omega = 0.5$). Non-integral values of y will often render the engineering problem physically meaningless and lead to severe numerical difficulties.

Androulakis [246] proposed a branch-and-bound algorithm for solving MIDO problems that arise when attempting to systematically reduce the complexity of kinetic mechanisms. This also requires the solution of dynamic optimization problems with relaxed binaries $\boldsymbol{\omega}$. Furthermore, as acknowledged by the author, branch-and-bound techniques are unsuitable for all but very small MIDO problems since they will typically involve the solution of a much larger number of dynamic optimization problems (the computational rate-determining step) than the decomposition methods described below.

The other MIDO solution approaches that have appeared in the literature are all based on decomposition principles, but differ in their treatment of the DAE system. One approach is to completely discretize the system using a technique such as orthogonal collocation on finite elements. This converts the MIDO problem Eq. 2.1 into a finite-dimensional, mixed-integer non-linear program (MINLP), where the continuous search variables for the optimization are the full set of discretization parameters for the differential, algebraic and control variables, and the time-invariant variables **d**. The MINLP can be solved using any conventional method. Balakrishna and Biegler [247] and Avraam et al. [248, 249] used the outer-approximation/equality-relaxation/augmented-penalty (OA/ER/AP) algorithm of Viswanathan and Grossmann [250], while Dimitriadis and Pistikopoulos [32] and Mohideen et al. [34] used generalised Benders decomposition (GBD) [251], in the contexts of synthesising reaction-separation systems, optimizing hybrid processes, analyzing dynamic systems under uncertainty, and simultaneous design and control, respectively. Schultz et al. [252] approximated the inequality path constraints by polynomials of second and third degrees and enforced constraints on the maximum values of these constraints. The advantages of these approaches are that the dynamic model and optimizer constraints are simultaneously converged and that general types of inequality constraints are easily handled. There are two major drawbacks, however. Firstly, the MINLP resulting from the discretization is very large even for relatively small problems. In the context of solving realistic engineering problems the size of the MINLP will often make the problem intractable. Secondly, intermediate solutions during the course of the optimization generally do not satisfy the original differential-algebraic system equations (hence the common use of the term "infeasible path" methods). This effectively renders the approach an "all-or-nothing" strategy since if the optimizer fails to find an optimum, the intermediate results are usually meaningless and cannot be used, e.g., as new initial starting points.

For continuous dynamic optimization problems, an alternative approach to complete discretization is to only parameterize the control variables, $\mathbf{u}(t)$, in terms of time-invariant parameters ("reduced space discretisation" or "control vector parameterization"). For given $\mathbf{u}(t)$ and values of the other search variables, $\mathbf{d}$, the DAE system is integrated using standard algorithms. Gradient information with respect to the objective function and constraints is obtained through finite difference perturbations, or more accurately through integration of the sensitivity equations (e.g. Vassiliadis, et al. [253, 254]), or solution of the adjoint equations (e.g. Sullivan and Sargent [244], Pytlak [255], Bloss et al. [256]). An NLP solver then adjusts the values of the controls and other search variables on the basis of these gradients. Chachuat et al. [257] presents an overview of these approaches. Although this method has the disadvantage that the solution of the DAEs and other equations can be computationally expensive, it gives useful intermediate (sub-optimal) solutions that are feasible with respect to the DAE system, allows more accurate control of the error associated with the solution of the DAE system, and is better conditioned when handling stiff models. As such, it is much more suited for large dynamic systems than the complete discretisation approach, although work is ongoing aimed at redressing the balance (e.g. Cervantes and Biegler [258]).

The MIDO algorithms in the literature that utilize reduced space methods (Allgor and Barton [259], Mohideen et al. [34, 36], Narraway [260], Schweiger and Floudas [261]; Sharif et al. [262]) all decompose the problem into a series of primal problems where the binary variables are fixed, and master problems which determine a new binary configuration for the next primal problem. In all cases the primal problem corresponds to a dynamic optimization problem that is solved using a reduced space approach, and which gives an upper bound on the final solution. The methods mainly differ in how they construct the master problem, which gives a lower bound on the solution. In the case of OA/ER/AP, the MILP master problem is constructed by linearizing the objective function and constraints. This contrasts with GBD where the master problem is constructed using the solution of the primal problem and corresponding dual information. The advantage of the former approach is that it requires the solution of fewer primal problems since its (more constrained) master problem always gives at least as tight a lower bound as the GBD master problem [263]. This could be important for MIDO problems since the primal problem (dynamic optimization) will tend to be much more computationally expensive relative to the master problem than for MINLPs. However, the OA methods are restricted to problems where the binary variables appear linearly and separably in the objective function and constraints, and are not designed to handle binary variables that appear explicitly in the DAE system (pointed out by Schweiger and Floudas [261]), which would be the case, e.g., for the simultaneous design and control optimization of a distillation column. Note that the convexity conditions required for OA/ER/AP and GBD to give the global optimum are rarely satisfied in dynamic process engineering problems. Allgor and Barton [259] proposed a different master problem approach where an MILP is solved based on a so-called "screening model". This approach may have the potential for providing rigorous bounds on the global optimum, its major drawback, however, is that the "screening model" is entirely case study-specific since it "is derived from domain specific knowledge gathered from physical laws and engineering insight".

Mohideen et al. [264] and Schweiger and Floudas [261] have both developed MIDO algorithms based on variant-2 of generalised Benders decomposition, v2-GBD [243], which are now explored in more detail. In both cases, the kth primal problem takes the form:

$$\begin{aligned}
\min_{\mathbf{u}(t),\mathbf{d},\boldsymbol{\omega}} \quad & J(\dot{\mathbf{x}}_d(t), \mathbf{x}_d(t), \mathbf{x}_a(t), \mathbf{u}(t), \mathbf{d}, \boldsymbol{\omega}^k, t) \\
s.t. \quad & \mathbf{h}_d(\dot{\mathbf{x}}_d(t), \mathbf{x}_d(t), \mathbf{x}_a(t), \mathbf{u}(t), \mathbf{d}, \boldsymbol{\omega}^k, t) = 0, \quad \forall t \in [t_0, t_f] \\
& \mathbf{h}_a(\mathbf{x}_d(t), \mathbf{x}_a(t), \mathbf{u}(t), \mathbf{d}, \boldsymbol{\omega}^k, t) = 0, \quad \forall t \in [t_0, t_f] \\
& \mathbf{h}_0(\dot{\mathbf{x}}_d(t_0), \mathbf{x}_d(t_0), \mathbf{x}_a(t_0), \mathbf{u}(t_0), \mathbf{d}, \boldsymbol{\omega}^k, t_0) = 0 \\
& \mathbf{h}_p(\dot{\mathbf{x}}_d(t_i), \mathbf{x}_d(t_i), \mathbf{x}_a(t_i), \mathbf{u}(t_i), \mathbf{d}, \boldsymbol{\omega}^k, t_i) = 0, \quad \forall t_i \in [t_0, t_f], i = 1, ..., N \\
& \mathbf{g}_p(\dot{\mathbf{x}}_d(t_i), \mathbf{x}_d(t_i), \mathbf{x}_a(t_i), \mathbf{u}(t_i), \mathbf{d}, \boldsymbol{\omega}^k, t_i) \leq 0, \quad \forall t_i \in [t_0, t_f], i = 1, ..., N \\
& \mathbf{h}_q(\mathbf{d}, \boldsymbol{\omega}^k) = 0, \\
& \mathbf{g}_q(\mathbf{d}, \boldsymbol{\omega}^k) \leq 0
\end{aligned} \tag{2.3}$$

where Eq. 2.3 corresponds to Eq. 2.1 with the binary variables fixed, $\boldsymbol{\omega} = \boldsymbol{\omega}^k$. In the approach of Mohideen et al. [264], the primal problem is solved using an implicit Runge–Kutta method to integrate the DAE system and the NLP gradient information is obtained using adjoint-based arguments. In the approach of Schweiger and Floudas [261], a backward difference formula algorithm is used for the DAEs and the sensitivity equations are integrated to obtain the required gradients. The solution of the kth primal problem is denoted as $\hat{\mathbf{u}}^k(t)$ and $\mathbf{d}^k$ with resulting differential variables $\hat{\mathbf{x}}_d^k(t)$ and algebraic variables $\hat{\mathbf{x}}_a^k(t)$.

The master problem is formulated using dual information and the solution of the primal problem in order to construct supporting functions about $\hat{\mathbf{u}}^k(t)$ and $\mathbf{d}^k$. If the primal problem is solved using adjoint-based arguments, as in Mohideen et al. [264], then the dual variables are available directly from the gradient evaluation method used for the primal problem. If a sensitivity-based approach is used for the primal problem then the dual variables are obtained through the solution of the final-value problem [265]:

$$\begin{aligned}
\frac{d}{dt}\left[\left(\frac{\partial \mathbf{h}_d}{\partial \dot{\mathbf{x}}_d}\right)^T \mathbf{v}_d^k(t)\right] - \left(\frac{\partial \mathbf{h}_d}{\partial \mathbf{x}_d}\right)^T \mathbf{v}_d^k(t) - \left(\frac{\partial \mathbf{h}_a}{\partial \dot{\mathbf{x}}_d}\right)^T \mathbf{v}_a^k(t) = 0 \\
-\left(\frac{\partial \mathbf{h}_d}{\partial \dot{\mathbf{x}}_a}\right)^T \mathbf{v}_d^k(t) - \left(\frac{\partial \mathbf{h}_a}{\partial \dot{\mathbf{x}}_a}\right)^T \mathbf{v}_a^k(t) = 0
\end{aligned} \tag{2.4}$$

with boundary conditions:

$$-\left(\frac{\partial J}{\partial \mathbf{x}_d}\right)^T + \left(\frac{\partial \mathbf{h}_d}{\partial \dot{\mathbf{x}}_d}\right)^T\Bigg|_{t_f} \mathbf{v}_d^k(t_f) + \left(\frac{\partial \mathbf{h}_d}{\partial \mathbf{x}_d}\right)^T\Bigg|_{t_f} \boldsymbol{\xi}_{d,f}^k + \left(\frac{\partial \mathbf{h}_a}{\partial \mathbf{x}_d}\right)^T\Bigg|_{t_f} \boldsymbol{\xi}_{a,f}^k + \left(\frac{\partial \mathbf{h}_p}{\partial \mathbf{x}_d}\right)^T \boldsymbol{\mu}_p^k$$
$$+\left(\frac{\partial \mathbf{g}_p}{\partial \mathbf{x}_d}\right)^T \boldsymbol{\lambda}_p^k = 0 \tag{2.5}$$

where $\mathbf{v}_d$, $\mathbf{v}_a$, μ, and λ are the dual multipliers associated with the differential equations, algebraic equations, equality point constraints and inequality point constraints, respectively. $\boldsymbol{\xi}_{d,f}$ and $\boldsymbol{\xi}_{a,f}$ are multipliers that are associated with the differential and algebraic equations at the final time, these are evaluated from the first order optimality conditions of the primal problem.

The dual multipliers $\boldsymbol{\mu}_k$ and $\boldsymbol{\lambda}_k$ are obtained from the solution of the primal problem, regardless of the method used to solve it. The Lagrange function, $\mathcal{L}^k$, for the case of a feasible primal problem Eq. 2.3 is given by:

$$\begin{aligned}
\mathcal{L}^k(\hat{\mathbf{u}}^k, \hat{\mathbf{d}}^k, \boldsymbol{\omega}) =& J\big(\hat{\dot{\mathbf{x}}}_d^k(t), \hat{\mathbf{x}}_d^k(t), \hat{\mathbf{x}}_a^k(t), \hat{\mathbf{u}}^k(t), \hat{\mathbf{d}}^k(t), \boldsymbol{\omega}, t\big) \\
&+ \int_{t_0}^{t_f} \mathbf{v}_d^{k^T}(t)\mathbf{h}_d\big(\hat{\dot{\mathbf{x}}}_d^k(t), \hat{\mathbf{x}}_d^k(t), \hat{\mathbf{x}}_a^k(t), \hat{\mathbf{u}}^k(t), \hat{\mathbf{d}}^k(t), \boldsymbol{\omega}, t\big)dt \\
&+ \int_{t_0}^{t_f} \mathbf{v}_a^{k^T}(t)\mathbf{h}_a\big(\hat{\mathbf{x}}_d^k(t), \hat{\mathbf{x}}_a^k(t), \hat{\mathbf{u}}^k(t), \hat{\mathbf{d}}^k(t), \boldsymbol{\omega}, t\big)dt \\
&+\boldsymbol{\xi}_{d,f}^{k\ T} \mathbf{h}_d\big(\hat{\dot{\mathbf{x}}}_d^k(t_f), \hat{\mathbf{x}}_d^k(t_f), \hat{\mathbf{x}}_a^k(t_f), \hat{\mathbf{u}}^k(t_f), \hat{\mathbf{d}}^k(t_f), \boldsymbol{\omega}, t_f\big)dt \\
&+\boldsymbol{\xi}_{a,f}^{k\ T} \mathbf{h}_d\big(\hat{\mathbf{x}}_d^k(t_f), \hat{\mathbf{x}}_a^k(t_f), \hat{\mathbf{u}}^k(t_f), \hat{\mathbf{d}}^k(t_f), \boldsymbol{\omega}, t_f\big)dt \\
&+\boldsymbol{\xi}_{d,0}^{k\ T} \mathbf{h}_d\big(\hat{\dot{\mathbf{x}}}_d^k(t_0), \hat{\mathbf{x}}_d^k(t_0), \hat{\mathbf{x}}_a^k(t_0), \hat{\mathbf{u}}^k(t_0), \hat{\mathbf{d}}^k(t_0), \boldsymbol{\omega}, t_0\big)dt \\
&+\boldsymbol{\xi}_{a,0}^{k\ T} \mathbf{h}_a\big(\hat{\mathbf{x}}_d^k(t_0), \hat{\mathbf{x}}_a^k(t_0), \hat{\mathbf{u}}^k(t_0), \hat{\mathbf{d}}^k(t_0), \boldsymbol{\omega}, t_0\big)dt \\
&+\boldsymbol{\kappa}^{k^T} \mathbf{h}_0\big(\hat{\dot{\mathbf{x}}}_d^k(t_0), \hat{\mathbf{x}}_d^k(t_0), \hat{\mathbf{x}}_a^k(t_0), \hat{\mathbf{u}}^k(t_0), \hat{\mathbf{d}}^k(t_0), \boldsymbol{\omega}, t_0\big)dt \\
&+\boldsymbol{\mu}_p^{k\ T} \mathbf{h}_p\big(\hat{\dot{\mathbf{x}}}_d^k(t_i), \hat{\mathbf{x}}_d^k(t_i), \hat{\mathbf{x}}_a^k(t_i), \hat{\mathbf{u}}^k(t_i), \hat{\mathbf{d}}^k(t_i), \boldsymbol{\omega}, t_i\big)dt \\
&+\boldsymbol{\lambda}_p^{k\ T} \mathbf{g}_p\big(\hat{\dot{\mathbf{x}}}_d^k(t_i), \hat{\mathbf{x}}_d^k(t_i), \hat{\mathbf{x}}_a^k(t_i), \hat{\mathbf{u}}^k(t_i), \hat{\mathbf{d}}^k(t_i), \boldsymbol{\omega}, t_i\big)dt \\
&+\boldsymbol{\mu}_q^{k\ T} \mathbf{h}_q(\hat{\mathbf{d}}^k, \boldsymbol{\omega}) + \boldsymbol{\lambda}_q^{k^T} \mathbf{g}_q(\hat{\mathbf{d}}^k, \boldsymbol{\omega})
\end{aligned} \tag{2.6}$$

where $\boldsymbol{\kappa}$ is the set of multipliers associated with the initial conditions, and $\boldsymbol{\xi}_{d,0}$ and $\boldsymbol{\xi}_{a,0}$ are the sets of multipliers associated with the initial values of the DAEs. These are all evaluated in a similar manner to $\boldsymbol{\xi}_{d,f}$ and $\boldsymbol{\xi}_{a,f}$.

The relaxed master problem is then:

$$\begin{aligned}
&\min_{\boldsymbol{\omega},\eta} \eta \\
&s.t.\quad \mathcal{L}^k(\hat{\mathbf{u}}, \hat{\mathbf{d}}, \boldsymbol{\omega}) \leq \eta, \quad k \in K_{feas}
\end{aligned} \tag{2.7}$$

where K_{feas} is the set of all feasible primal problems solved up to the iteration under consideration. If the variables $\boldsymbol{\omega}$ participate linearly in Eq. 2.1, then Eq. 2.7 is an MILP that can be solved via conventional branch and bound methods. Note that integer cuts of the form [243]:

$$\begin{aligned} &\sum_{i \in B^k} \omega_i - \sum_{i \in NB^k} \omega_i \leq |B^k| - 1 \\ &B^k = \{i : \omega_i^k = 1\} \\ &NB^k = \{i : \omega_i^k = 0\} \end{aligned} \tag{2.8}$$

where $|B^k|$ is the cardinality of B^k, are typically introduced in Eq. 2.7 in order to exclude previous integer solutions. Pure binary constraints on the original system are also included in Eq. 2.7.

The MIDO algorithm terminates when the difference between the least upper bound from the primal problems and the lower bound from the master problem is less than a specified tolerance, or if there is an infeasible master problem, in which case all feasible integer combinations have been considered. The solution to the MIDO problem then corresponds to the solution of the primal problem with the least upper bound.

2.4 An Alternative Approach

In this section an alternative MIDO approach is proposed, still based on v2-GBD principles, but which does not require the computation of the dual variables via the complex integration in Eq. 2.4, for example, and which leads to the formulation of an equivalent but much more compact master problem than Eqs. 2.6 and 2.7. In order to achieve this, a modification of problem Eq. 2.1 is considered where the original binary variables, $\boldsymbol{\omega}$, are relaxed, but are still forced to take integral values by the addition of new binary variables, $\bar{\boldsymbol{\omega}}$, and constraints $\mathbf{h}_{\omega} = \boldsymbol{\omega} - \bar{\boldsymbol{\omega}} = 0$.

2.4.1 Primal Problem

At the kth iteration of the GBD algorithm, the following dynamic optimization problem is solved:

$$\begin{aligned} \min_{\mathbf{u}(t),\mathbf{d},\boldsymbol{\omega}} \quad & J(\dot{\mathbf{x}}_d(t_f), \mathbf{x}_d(t_f), \mathbf{x}_a(t_f), \mathbf{u}(t_f), \mathbf{d}, \boldsymbol{\omega}, t_f) \\ s.t. \quad & \mathbf{h}_d(\dot{\mathbf{x}}_d(t), \mathbf{x}_d(t), \mathbf{x}_a(t), \mathbf{u}(t), \mathbf{d}, \boldsymbol{\omega}, t) = 0, \quad \forall t \in [t_0, t_f] \\ & \mathbf{h}_a(\mathbf{x}_d(t), \mathbf{x}_a(t), \mathbf{u}(t), \mathbf{d}, \boldsymbol{\omega}, t) = 0, \quad \forall t \in [t_0, t_f] \\ & \mathbf{h}_0(\dot{\mathbf{x}}_d(t_0), \mathbf{x}_d(t_0), \mathbf{x}_a(t_0), \mathbf{u}(t_0), \mathbf{d}, \boldsymbol{\omega}, t_0) = 0 \end{aligned}$$

$$
\begin{aligned}
&\mathbf{h}_p(\dot{\mathbf{x}}_d(t_i), \mathbf{x}_d(t_i), \mathbf{x}_a(t_i), \mathbf{u}(t_i), \mathbf{d}, \boldsymbol{\omega}, t_i) = 0, \quad \forall t_i \in [t_0, t_f], i = 1, ..., N \\
&\mathbf{g}_p(\dot{\mathbf{x}}_d(t_i), \mathbf{x}_d(t_i), \mathbf{x}_a(t_i), \mathbf{u}(t_i), \mathbf{d}, \boldsymbol{\omega}, t_i) \leq 0, \quad \forall t_i \in [t_0, t_f], i = 1, ..., N \\
&\mathbf{h}_q(\mathbf{d}, \boldsymbol{\omega}) = 0, \\
&\mathbf{g}_q(\mathbf{d}, \boldsymbol{\omega}) \leq 0 \\
&\mathbf{h}_{\omega} = \boldsymbol{\omega} - \bar{\boldsymbol{\omega}} = 0
\end{aligned}
\tag{2.9}
$$

where $\boldsymbol{\omega}$ is now a set of continuous, time-invariant search variables and $\bar{\boldsymbol{\omega}}$ is the set of complicating (binary) variables. If the same values are chosen for the complicating variables in problems (2.3) and (2.9), then these two problems are clearly identical. Thus, their solutions are both denoted by $\hat{\mathbf{u}}^k(t)$ and $\hat{\mathbf{d}}^k$, with resulting state variables $\hat{\mathbf{x}}_d^k(t)$ and algebraic variables $\hat{\mathbf{x}}_a^k(t)$. The vector $\boldsymbol{\omega}$ at the optimum solution to Eq. 2.9 is given by $\boldsymbol{\omega}^k$.

2.4.2 Construction of the Master Problem

As before, the GBD relaxed master problem is formulated using dual information from the solution of the primal problem. It is obvious that the dual variables associated with $\mathbf{h}_d$, $\mathbf{h}_a$, $\mathbf{h}_0$, $\mathbf{h}_p$, $\mathbf{g}_p$, $\mathbf{h}_q$, and $\mathbf{g}_q$ are identical for the "old" and "new" primal formulations (2.3) and (2.9). For a feasible primal problem Eq. 2.9, the Lagrange function, $\bar{\mathcal{L}}^k$, is given by:

$$
\begin{aligned}
\mathcal{L}^k(\hat{\mathbf{u}}^k, \hat{\mathbf{d}}^k, \bar{\boldsymbol{\omega}}) =& J\big(\hat{\dot{\mathbf{x}}}_d^k(t_f), \hat{\mathbf{x}}_d^k(t_f), \hat{\mathbf{x}}_a^k(t_f), \hat{\mathbf{u}}^k(t_f), \hat{\mathbf{d}}^k(t_f), \boldsymbol{\omega}^k, t_f\big) \\
&+ \int_{t_0}^{t_f} \mathbf{v}_d^{k^T}(t)\mathbf{h}_d\big(\hat{\dot{\mathbf{x}}}_d^k(t), \hat{\mathbf{x}}_d^k(t), \hat{\mathbf{x}}_a^k(t), \hat{\mathbf{u}}^k(t), \hat{\mathbf{d}}^k(t), \boldsymbol{\omega}^k, t\big)dt \\
&+ \int_{t_0}^{t_f} \mathbf{v}_a^{k^T}(t)\mathbf{h}_a\big(\hat{\mathbf{x}}_d^k(t), \hat{\mathbf{x}}_a^k(t), \hat{\mathbf{u}}^k(t), \hat{\mathbf{d}}^k(t), \boldsymbol{\omega}^k, t\big)dt \\
&+ \boldsymbol{\xi}_{d,f}^{k\ T}\mathbf{h}_d\big(\hat{\dot{\mathbf{x}}}_d^k(t_f), \hat{\mathbf{x}}_d^k(t_f), \hat{\mathbf{x}}_a^k(t_f), \hat{\mathbf{u}}^k(t_f), \hat{\mathbf{d}}^k(t_f), \boldsymbol{\omega}^k, t_f\big)dt \\
&+ \boldsymbol{\xi}_{a,f}^{k\ T}\mathbf{h}_d\big(\hat{\mathbf{x}}_d^k(t_f), \hat{\mathbf{x}}_a^k(t_f), \hat{\mathbf{u}}^k(t_f), \hat{\mathbf{d}}^k(t_f), \boldsymbol{\omega}^k, t_f\big)dt \\
&+ \boldsymbol{\xi}_{d,0}^{k\ T}\mathbf{h}_d\big(\hat{\dot{\mathbf{x}}}_d^k(t_0), \hat{\mathbf{x}}_d^k(t_0), \hat{\mathbf{x}}_a^k(t_0), \hat{\mathbf{u}}^k(t_0), \hat{\mathbf{d}}^k(t_0), \boldsymbol{\omega}^k, t_0\big)dt \\
&+ \boldsymbol{\xi}_{a,0}^{k\ T}\mathbf{h}_a\big(\hat{\mathbf{x}}_d^k(t_0), \hat{\mathbf{x}}_a^k(t_0), \hat{\mathbf{u}}^k(t_0), \hat{\mathbf{d}}^k(t_0), \boldsymbol{\omega}^k, t_0\big)dt \\
&+ \boldsymbol{\kappa}^{k^T}\mathbf{h}_0\big(\hat{\dot{\mathbf{x}}}_d^k(t_0), \hat{\mathbf{x}}_d^k(t_0), \hat{\mathbf{x}}_a^k(t_0), \hat{\mathbf{u}}^k(t_0), \hat{\mathbf{d}}^k(t_0), \boldsymbol{\omega}^k, t_0\big)dt \\
&+ \boldsymbol{\mu}_p^{k\ T}\mathbf{h}_p\big(\hat{\dot{\mathbf{x}}}_d^k(t_i), \hat{\mathbf{x}}_d^k(t_i), \hat{\mathbf{x}}_a^k(t_i), \hat{\mathbf{u}}^k(t_i), \hat{\mathbf{d}}^k(t_i), \boldsymbol{\omega}^k, t_i\big)dt \\
&+ \boldsymbol{\lambda}_p^{k\ T}\mathbf{g}_p\big(\hat{\dot{\mathbf{x}}}_d^k(t_i), \hat{\mathbf{x}}_d^k(t_i), \hat{\mathbf{x}}_a^k(t_i), \hat{\mathbf{u}}^k(t_i), \hat{\mathbf{d}}^k(t_i), \boldsymbol{\omega}^k, t_i\big)dt \\
&+ \boldsymbol{\mu}_q^{k^T}\mathbf{h}_q(\hat{\mathbf{d}}^k, \boldsymbol{\omega}^k) + \boldsymbol{\lambda}_q^{k^T}\mathbf{g}_q(\hat{\mathbf{d}}^k, \boldsymbol{\omega}^k) \\
&+ \boldsymbol{\zeta}^{k^T}(\boldsymbol{\omega}_k - \bar{\boldsymbol{\omega}})
\end{aligned}
\tag{2.10}
$$

where $\boldsymbol{\zeta}$ is the set of dual multipliers for the equations $\mathbf{h}_\omega = 0$. Since $\mathbf{h}_d$, $\mathbf{h}_a$, $\mathbf{h}_0$, $\mathbf{h}_p$, and $\mathbf{h}_q$ all equal 0 at the primal solution and they are not functions of the complicating variables $\bar{\boldsymbol{\omega}}$, their associated terms can be removed from the Lagrangian in Eq. 2.10. Similarly, the complementarity optimality conditions for the NLP in the primal problem ensure that $\boldsymbol{\lambda}_p^T \mathbf{g}_p = \boldsymbol{\lambda}_p^T \mathbf{g}_q = 0$, and since $\mathbf{g}_p$ and $\mathbf{g}_q$ are not functions of $\bar{\boldsymbol{\omega}}$ either, their associated terms can also be removed. Hence, the Lagrange function $\bar{\mathcal{L}}^k$ can be simplified to:

$$\begin{aligned}\mathcal{L}^k(\hat{\mathbf{u}}^k, \hat{\mathbf{d}}^k, \bar{\boldsymbol{\omega}}) =& J\left(\hat{\mathbf{x}}_d^k(t_f), \hat{\mathbf{x}}_d^k(t_f), \hat{\mathbf{x}}_a^k(t_f), \hat{\mathbf{u}}^k(t_f), \hat{\mathbf{d}}^k(t_f), \boldsymbol{\omega}^k, t_f\right) \\ &+ {\boldsymbol{\zeta}^k}^T(\boldsymbol{\omega}_k - \bar{\boldsymbol{\omega}})\end{aligned} \tag{2.11}$$

and the master problem posed as:

$$\begin{aligned}&\min_{\bar{\boldsymbol{\omega}}, \eta} \eta \\ &s.t. \quad \bar{\mathcal{L}}^k(\hat{\mathbf{u}}, \hat{\mathbf{d}}, \bar{\boldsymbol{\omega}}) \le \eta, \quad k \in K_{feas}\end{aligned} \tag{2.12}$$

It is clear from the Lagrange function definition in Eq. 2.11 that the master problem Eq. 2.12 does not require any dual information with respect to the DAE system and so no intermediate adjoint problem is required for its construction. It should also be noted that the master problem Eq. 2.12 is exactly equivalent to the master problem Eq. 2.7. A proof of this can be found in Appendix A.1.

2.5 New Algorithm I

Based on the theoretical developments presented so far, a new, generalized Benders decomposition-based algorithm for the solution of the general MIDO problem Eq. 2.1 can be summarized as follows:

Step 1. Set the termination tolerance, ϵ. Initialize the iteration counter, $k = 1$, lower bound, $LB = -\infty$, and upper bound, $UB = \infty$.

Step 2. For fixed values of the binary variables, $\boldsymbol{\omega} = \boldsymbol{\omega}^k$, solve the kth primal problem Eq. 2.3 to obtain a solution J^k. (For computational speed, it may be useful to omit the search variables and constraints that are superfluous due to the current choice of values for the binary variables.) Set $UB = min(UB, J^k)$ and store the values of the continuous and binary variables corresponding to the best primal solution so far.

Step 3. Solve the primal problem Eq. 2.9 at the solution found in Step 2 with the full set of search variables and constraints included. Convergence will be achieved in one iteration. Obtain the multipliers $\boldsymbol{\imath}^k$ needed to construct the kth master problem Eq. 2.12.

Step 4. Solve the kth relaxed master problem Eq. 2.12 with integer cuts Eq. 2.8 and the pure binary constraints of the original system, to obtain a solution η^k. Update the lower bound, $LB = \eta^k$.

Step 5. If $UB - LB \leq \epsilon$, or the master problem is infeasible, then stop. The optimal solution corresponds to UB and the values stored in Step 2. Otherwise, set $k = k + 1$ and $\boldsymbol{\omega}^k + 1$ equal to the integer solution of the kth master problem, and return to Step 2.

Note that if in Step 2 a particular set of values $\boldsymbol{\omega}^k$ renders the primal problem Eq. 2.3 infeasible, then an infeasibility minimization problem is solved instead. This can involve, for example, minimizing the l_1 or l_∞ sum of constraint violations (Floudas, 1995). In Step 3, the infeasibility minimization problem is then resolved at its optimal solution to obtain multipliers $\boldsymbol{\zeta}^k$. The formulation corresponding to minimizing the l_∞ sum of violations is shown below:

$$
\begin{aligned}
\min_{\alpha,\mathbf{u}(t),\mathbf{d},\boldsymbol{\omega}} \quad & \alpha \\
s.t. \quad & \mathbf{h}_d(\dot{\mathbf{x}}_d(t), \mathbf{x}_d(t), \mathbf{x}_a(t), \mathbf{u}(t), \mathbf{d}, \boldsymbol{\omega}, t) = 0, \quad \forall t \in [t_0, t_f] \\
& \mathbf{h}_a(\mathbf{x}_d(t), \mathbf{x}_a(t), \mathbf{u}(t), \mathbf{d}, \boldsymbol{\omega}, t) = 0, \quad \forall t \in [t_0, t_f] \\
& \mathbf{h}_0(\dot{\mathbf{x}}_d(t_0), \mathbf{x}_d(t_0), \mathbf{x}_a(t_0), \mathbf{u}(t_0), \mathbf{d}, \boldsymbol{\omega}, t_0) = 0 \\
& \mathbf{h}_p(\dot{\mathbf{x}}_d(t_i), \mathbf{x}_d(t_i), \mathbf{x}_a(t_i), \mathbf{u}(t_i), \mathbf{d}, \boldsymbol{\omega}, t_i) = 0, \quad \forall t_i \in [t_0, t_f], i = 1, ..., N \\
& \mathbf{g}_p(\dot{\mathbf{x}}_d(t_i), \mathbf{x}_d(t_i), \mathbf{x}_a(t_i), \mathbf{u}(t_i), \mathbf{d}, \boldsymbol{\omega}^k, t_i) \leq 0, \quad \forall t_i \in [t_0, t_f], i = 1, ..., N \\
& \mathbf{h}_q(\mathbf{d}, \boldsymbol{\omega}) = 0, \\
& \mathbf{g}_q(\mathbf{d}, \boldsymbol{\omega}) \leq 0 \\
& \mathbf{h}_\omega = \boldsymbol{\omega} - \bar{\boldsymbol{\omega}} = 0 \\
& \alpha \leq 0
\end{aligned}
\tag{2.13}
$$

where α is a positive scalar quantity and $\mathbf{e}$ is a vector whose elements are all equal to one. In Step 4, the master problem Eq. 2.12 is then augmented as:

$$
\begin{aligned}
\min_{\bar{\boldsymbol{\omega}},\eta} \quad & \eta \\
s.t. \quad & J\left(\hat{\dot{\mathbf{x}}}_d^k(t_f), \hat{\mathbf{x}}_d^k(t_f), \hat{\mathbf{x}}_a^k(t_f), \hat{\mathbf{u}}^k(t_f), \hat{\mathbf{d}}^k(t_f), \boldsymbol{\omega}^k, t_f\right) + \boldsymbol{\zeta}^{k^T}(\boldsymbol{\omega}_k - \bar{\boldsymbol{\omega}}) \leq \eta \\
& \boldsymbol{\zeta}^{k^T}(\boldsymbol{\omega}_k - \bar{\boldsymbol{\omega}}), \quad k \in K_{feas}
\end{aligned}
\tag{2.14}
$$

where K_{feas} denotes the set of iteration numbers for which feasible primal problems are found and K_{infeas} the set for which infeasible primal problems are found.

2.6 Remarks on MIDO Algorithm I

(i) As already stated, one advantage of the algorithm outlined above is that, even when the binary variables $\boldsymbol{\omega}$ appear explicitly (and nonlinearly) in the DAE system, the master problem Eq. 2.12 does not require any dual information with respect to the DAE system and so no intermediate adjoint problem is required for its construction. This comes at the expense of adding more equations, $\mathbf{h}_y = 0$, and search variables, $\bar{\boldsymbol{\omega}}$. However, these are only used in Step 3 when the primal problem is re-solved at its optimal solution, and convergence is achieved in just one iteration. The use of this novel strategy also circumvents any numerical difficulties (such as those suggested by Schweiger and Floudas [261]) associated with reformulating the primal problem.

(ii) The master problem in Step 4 of the new MIDO algorithm has a much simpler form compared to algorithms where adjoint variables are required. In the latter case, the master problem defined by Eqs. 2.6 and 2.7, has the form:

$$
\begin{aligned}
&\min_{\boldsymbol{\omega},\eta} \quad \eta \\
&s.t. \quad J\big(\hat{\dot{\mathbf{x}}}_d^k(t_f), \hat{\mathbf{x}}_d^k(t_f), \hat{\mathbf{x}}_a^k(t_f), \hat{\mathbf{u}}^k(t_f), \hat{\mathbf{d}}^k(t_f), \boldsymbol{\omega}, t_f\big) \\
&\quad + \int_{t_0}^{t_f} \mathbf{v}_d^{k^T}(t)\mathbf{h}_d\big(\hat{\dot{\mathbf{x}}}_d^k(t), \hat{\mathbf{x}}_d^k(t), \hat{\mathbf{x}}_a^k(t), \hat{\mathbf{u}}^k(t), \hat{\mathbf{d}}^k(t), \boldsymbol{\omega}, t\big)dt \\
&\quad + \int_{t_0}^{t_f} \mathbf{v}_a^{k^T}(t)\mathbf{h}_a\big(\hat{\mathbf{x}}_d^k(t), \hat{\mathbf{x}}_a^k(t), \hat{\mathbf{u}}^k(t), \hat{\mathbf{d}}^k(t), \boldsymbol{\omega}^k, t\big)dt \\
&\quad + \boldsymbol{\xi}_{d,f}^{k\;T}\mathbf{h}_d\big(\hat{\dot{\mathbf{x}}}_d^k(t_f), \hat{\mathbf{x}}_d^k(t_f), \hat{\mathbf{x}}_a^k(t_f), \hat{\mathbf{u}}^k(t_f), \hat{\mathbf{d}}^k(t_f), \boldsymbol{\omega}, t_f\big)dt \\
&\quad + \boldsymbol{\xi}_{a,f}^{k\;T}\mathbf{h}_d\big(\hat{\mathbf{x}}_d^k(t_f), \hat{\mathbf{x}}_a^k(t_f), \hat{\mathbf{u}}^k(t_f), \hat{\mathbf{d}}^k(t_f), \boldsymbol{\omega}, t_f\big)dt \\
&\quad + \boldsymbol{\xi}_{d,0}^{k\;T}\mathbf{h}_d\big(\hat{\dot{\mathbf{x}}}_d^k(t_0), \hat{\mathbf{x}}_d^k(t_0), \hat{\mathbf{x}}_a^k(t_0), \hat{\mathbf{u}}^k(t_0), \hat{\mathbf{d}}^k(t_0), \boldsymbol{\omega}, t_0\big)dt \\
&\quad + \boldsymbol{\xi}_{a,0}^{k\;T}\mathbf{h}_a\big(\hat{\mathbf{x}}_d^k(t_0), \hat{\mathbf{x}}_a^k(t_0), \hat{\mathbf{u}}^k(t_0), \hat{\mathbf{d}}^k(t_0), \boldsymbol{\omega}, t_0\big)dt \\
&\quad + \boldsymbol{\kappa}^{k^T}\mathbf{h}_0\big(\hat{\dot{\mathbf{x}}}_d^k(t_0), \hat{\mathbf{x}}_d^k(t_0), \hat{\mathbf{x}}_a^k(t_0), \hat{\mathbf{u}}^k(t_0), \hat{\mathbf{d}}^k(t_0), \boldsymbol{\omega}, t_0\big)dt \\
&\quad + \boldsymbol{\mu}_p^{k\;T}\mathbf{h}_p\big(\hat{\dot{\mathbf{x}}}_d^k(t_i), \hat{\mathbf{x}}_d^k(t_i), \hat{\mathbf{x}}_a^k(t_i), \hat{\mathbf{u}}^k(t_i), \hat{\mathbf{d}}^k(t_i), \boldsymbol{\omega}, t_i\big)dt \\
&\quad + \boldsymbol{\lambda}_p^{k\;T}\mathbf{g}_p\big(\hat{\dot{\mathbf{x}}}_d^k(t_i), \hat{\mathbf{x}}_d^k(t_i), \hat{\mathbf{x}}_a^k(t_i), \hat{\mathbf{u}}^k(t_i), \hat{\mathbf{d}}^k(t_i), \boldsymbol{\omega}, t_i\big)dt \\
&\quad + \boldsymbol{\mu}_q^{k\;T}\mathbf{h}_q(\hat{\mathbf{d}}^k, \boldsymbol{\omega}) + \boldsymbol{\lambda}_q^{k\;T}\mathbf{g}_q(\hat{\mathbf{d}}^k, \boldsymbol{\omega}) \\
&\quad \leq \eta
\end{aligned}
\tag{2.15}
$$

Construction of this master problem thus requires that the objective function and all the DAEs, equality and inequality constraints that involve the binary variables $\boldsymbol{\omega}$ be considered together with their adjoint and Lagrange multipliers. Their number may be considerably larger than the number of binary variables $\boldsymbol{\omega}$, especially for complex models. In contrast, the master problem defined by Eqs. 2.11 and 2.12, for the new MIDO algorithm is:

$$\begin{aligned} &\min_{\omega,\eta} \quad \eta \\ &s.t. \quad J\left(\hat{\dot{\mathbf{x}}}_d^k(t_f), \hat{\mathbf{x}}_d^k(t_f), \hat{\mathbf{x}}_a^k(t_f), \hat{\mathbf{u}}^k(t_f), \hat{\mathbf{d}}^k(t_f), \omega^k, t_f\right) + {\zeta^k}^T(\omega^k - \bar{\omega}) \end{aligned} \tag{2.16}$$

which only requires the value of the primal objective function together with extra terms whose complexity is independent of the complexity of the process model. The practical benefit of this will be seen in Example 2, presented later in this chapter.

(iii) Another important benefit of the new MIDO algorithm is that it is independent of the type of method used for solving the dynamic optimization primal problem. The upshot of this is that with the formulation presented, any existing dynamic optimization solver can be used to solve MIDO problems since even the most elementary NLP codes give the simple Lagrange multiplier information required for the master problem. Note, however, that the advantages gained from the new algorithm do depend on the method used for the primal solution. The maximum benefits are obtained if control vector parameterization is used since no evaluation of the adjoint variables is then necessary. If a multiple shooting approach is used, then the adjoints are already computed at the primal solution; however, the master problem formulated with Algorithm I is still much simpler and more attractive from a computational viewpoint than the master problem that would result otherwise.

(iv) Since the new MIDO algorithm is based on v2-GBD, like the MIDO approaches of Mohideen et al. [264] and Schweiger and Floudas [261], it shares its limitations. In particular, the algorithm is only guaranteed to converge to the global optimum under convexity conditions for the primal problem with the binary variables appearing linearly and separably [243]. In the much more common, nonconvex case, there is a possibility that the GBD relaxed master problems Eqs. 2.7 and 2.12, will exclude potentially feasible choices of $\boldsymbol{\omega}$ from the solution set.

Note that for (MI)NLP problems, it has been shown that GBD can lead to solutions which are not even local optima if convexity criteria are not satisfied and a set of continuous variables is chosen as the complicating (fixed) variables for the primal problem [266–268]. Perhaps in response to this, Barton, Allgor, Feehery and Galan [269] and Allgor and Barton [259] incorrectly suggested that the use of GBD approaches for MIDO "has the property of convergence to an arbitrary point which is not even guaranteed to be a local minimum." In fact, the new MIDO algorithm (and the GBD algorithms of Mohideen et al. [264] and Schweiger and Floudas [261]) will always converge to at least a local optimum, where [236]:

Definition 1 A local optimum of an MINLP is one which, for fixed binaries $\boldsymbol{\omega}$, satisfies local optimality conditions for the resulting (primal) problem.

The above definition is intuitively obvious since when $\boldsymbol{\omega}$ is discrete there can be no feasible neighbourhood around it which improves the objective. Therefore, the (local optimal) solution of any of the feasible primal problems will result in a local optimum of the MIDO problem.

2.7 Extending Algorithm I to Handle Time-Dependent Binary Variables

Ross et al. [236] showed that the GBD-based MIDO algorithm of Mohideen et al. [264] can be readily extended to tackle MIDO problems which involve time-dependent binary variables. Such problems can arise in areas such as batch scheduling and adaptive control, where in the latter, the control structure changes with time. The new MIDO algorithm presented in this article can also tackle problems involving time-dependent binary variables, which will be denoted by $\mathbf{u}_y(t)$. The general formulation of such a problem is:

$$
\begin{aligned}
\min_{\mathbf{u}(t),\mathbf{d},\boldsymbol{\omega},\mathbf{u}_y(t)} \quad & J(\dot{\mathbf{x}}_d(t_f), \mathbf{x}_d(t_f), \mathbf{x}_a(t_f), \mathbf{u}(t_f), \mathbf{d}, \boldsymbol{\omega}, \mathbf{u}_y(t), t_f) \\
s.t. \quad & \mathbf{h}_d(\dot{\mathbf{x}}_d(t), \mathbf{x}_d(t), \mathbf{x}_a(t), \mathbf{u}(t), \mathbf{d}, \boldsymbol{\omega}, \mathbf{u}_y(t), t) = 0, \quad \forall t \in [t_0, t_f] \\
& \mathbf{h}_a(\mathbf{x}_d(t), \mathbf{x}_a(t), \mathbf{u}(t), \mathbf{d}, \boldsymbol{\omega}, \mathbf{u}_y(t), t) = 0, \quad \forall t \in [t_0, t_f] \\
& \mathbf{h}_0(\dot{\mathbf{x}}_d(t_0), \mathbf{x}_d(t_0), \mathbf{x}_a(t_0), \mathbf{u}(t_0), \mathbf{d}, \boldsymbol{\omega}, \mathbf{u}_y(t_0), t_0) = 0 \\
& \mathbf{h}_p(\dot{\mathbf{x}}_d(t_i), \mathbf{x}_d(t_i), \mathbf{x}_a(t_i), \mathbf{u}(t_i), \mathbf{d}, \boldsymbol{\omega}, \mathbf{u}_y(t_i), t_i) = 0, \quad \forall t_i \in [t_0, t_f], i = 1, ..., N \\
& \mathbf{g}_p(\dot{\mathbf{x}}_d(t_i), \mathbf{x}_d(t_i), \mathbf{x}_a(t_i), \mathbf{u}(t_i), \mathbf{d}, \boldsymbol{\omega}, \mathbf{u}_y(t_i), t_i) \le 0, \quad \forall t_i \in [t_0, t_f], i = 1, ..., N \\
& \mathbf{h}_q(\mathbf{d}, \boldsymbol{\omega}) = 0 \\
& \mathbf{g}_q(\mathbf{d}, \boldsymbol{\omega}) \le 0 \\
& \mathbf{h}_\omega = \boldsymbol{\omega} - \bar{\boldsymbol{\omega}} = 0
\end{aligned}
\tag{2.17}
$$

$\mathbf{u}_y(t)$ may be parameterized by means of a piecewise-constant profile defined on N_{uy} sub-intervals over the time horizon $[t_0, t_f]$. Hence:

$$
\mathbf{u}_y(t) = \mathbf{u}_y(j), \quad j = 1, ..., N_{uy} \tag{2.18}
$$

The values of $\mathbf{u}_y(t)$ during the jth sub-interval are given by a set of time-invariant binaries $\mathbf{u}_y(j)$. With these binaries fixed, Eq. 2.17 becomes a standard dynamic optimization problem. A similar approach to that described earlier can thus be employed. The primal problem in Step 3 of Algorithm I becomes:

$$
\begin{aligned}
\min_{\mathbf{u}(t),\mathbf{d},\boldsymbol{\omega},\mathbf{u}_{y(j)}} \quad & J(\dot{\mathbf{x}}_d(t_f), \mathbf{x}_d(t_f), \mathbf{x}_a(t_f), \mathbf{u}(t_f), \mathbf{d}, \boldsymbol{\omega}, \mathbf{u}_y(t), t_f) \\
s.t. \quad & \mathbf{h}_d(\dot{\mathbf{x}}_d(t), \mathbf{x}_d(t), \mathbf{x}_a(t), \mathbf{u}(t), \mathbf{d}, \boldsymbol{\omega}, \mathbf{u}_y(t), t) = 0, \quad \forall t \in [t_0, t_f] \\
& \mathbf{h}_a(\mathbf{x}_d(t), \mathbf{x}_a(t), \mathbf{u}(t), \mathbf{d}, \boldsymbol{\omega}, \mathbf{u}_y(t), t) = 0, \quad \forall t \in [t_0, t_f] \\
& \mathbf{h}_0(\dot{\mathbf{x}}_d(t_0), \mathbf{x}_d(t_0), \mathbf{x}_a(t_0), \mathbf{u}(t_0), \mathbf{d}, \boldsymbol{\omega}, \mathbf{u}_y(t_0), t_0) = 0 \\
& \mathbf{h}_p(\dot{\mathbf{x}}_d(t_i), \mathbf{x}_d(t_i), \mathbf{x}_a(t_i), \mathbf{u}(t_i), \mathbf{d}, \boldsymbol{\omega}, \mathbf{u}_y(t_i), t_i) = 0, \quad \forall t_i \in [t_0, t_f], i = 1, ..., N \\
& \mathbf{g}_p(\dot{\mathbf{x}}_d(t_i), \mathbf{x}_d(t_i), \mathbf{x}_a(t_i), \mathbf{u}(t_i), \mathbf{d}, \boldsymbol{\omega}, \mathbf{u}_y(t_i), t_i) \leq 0, \quad \forall t_i \in [t_0, t_f], i = 1, ..., N \\
& \mathbf{h}_q(\mathbf{d}, \boldsymbol{\omega}) = 0 \\
& \mathbf{g}_q(\mathbf{d}, \boldsymbol{\omega}) \leq 0 \\
& \mathbf{h}_\omega = \boldsymbol{\omega} - \bar{\boldsymbol{\omega}} = 0 \\
& \mathbf{h}_{\mathbf{u}_y(j)} = \mathbf{u}_{y(j)} - \bar{\mathbf{u}}^k_{y(j)} = 0, \quad j = 1, ..., N_{uy}
\end{aligned}
\tag{2.19}
$$

In Eq. 2.19 $\mathbf{u}_y(j)$, $j = 1, ..., N_{uy}$ and $\boldsymbol{\omega}$ are continuous search variables, while $\bar{\mathbf{u}}_y(j)$, $j = 1, ..., N_{uy}$ and $\bar{\boldsymbol{\omega}}$ are the complicating, binary variables. The constraints $\mathbf{h}_{\mathbf{u}_y}(j) = 0$ are interior-point constraints that are handled in a similar fashion to the point constraints $\mathbf{h}_p = 0$. The multipliers associated with $\mathbf{h}_{\mathbf{u}_y}(j)$ are denoted by $\boldsymbol{\zeta}_{\mathbf{u}_y}(j)$. The values of the variables at the optimum are denoted as before with the addition of $\hat{\mathbf{u}}^k_y(j)$. The master problem is formulated in a similar manner to Eq. 2.14:

$$
\begin{aligned}
\min_{\bar{\boldsymbol{\omega}},\bar{\mathbf{u}}_{y(j)},\eta} \quad & \eta \\
s.t. \quad & J^k + \boldsymbol{\zeta}^{k^T}(\boldsymbol{\omega}^k - \bar{\boldsymbol{\omega}}) + \sum_{j=1}^{N_{uy}} [\boldsymbol{\zeta}^k_{\mathbf{u}_y}(j)]^T [\hat{\mathbf{u}}^k_y(j) - \bar{\mathbf{u}}_y(j)] \leq \eta, \quad k \in K_{feas} \\
& \boldsymbol{\zeta}^{k^T}(\boldsymbol{\omega}^k - \bar{\boldsymbol{\omega}}) + \sum_{j=1}^{N_{uy}} [\boldsymbol{\zeta}^k_{\mathbf{u}_y}(j)]^T [\hat{\mathbf{u}}^k_y(j) - \bar{\mathbf{u}}_y(j)] \leq 0, \quad k \in K_{feas}
\end{aligned}
\tag{2.20}
$$

2.8 Using Outer-Approximation to Construct the Master Problem

As discussed in the Sect. 2.3 earlier in this chapter, current MIDO algorithms that use the principles of OA/ER to construct the master problem are limited in that they are not applicable to models where the binary variables appear explicitly in the DAE system. This can be overcome using the reformulation strategy that was applied in Algorithm I.

In an OA/ER-based MIDO algorithm, the primal problems are solved in an identical manner to those in Step 2 and Step 3 of the GBD-based MIDO Algorithm I. However, in order to construct the master problems, instead of obtaining Lagrange multiplier values in Step 3, gradient information is required of the objective function and constraints with respect to the optimization search variables. During solution of the primal problem, the independent search variables are $\mathbf{d}$, $\boldsymbol{\omega}$ and the time-invariant variables resulting from the parameterization of the controls, which are denoted by $\mathbf{u}_j$. The kth master problem (for the case of feasible primal problems) is thus:

$$
\begin{aligned}
&\min_{\bar{\mathbf{u}},\mathbf{d},\boldsymbol{\omega},\mathbf{u}_j,\eta} \quad \eta \\
&s.t. \quad \eta \geq J^k + \left(\frac{\partial J}{\partial \mathbf{d}}\right)\bigg|_k (\mathbf{d}-\hat{\mathbf{d}}^k) + \left(\frac{\partial J}{\partial \boldsymbol{\omega}}\right)\bigg|_k (\boldsymbol{\omega}-\hat{\boldsymbol{\omega}}^k) + \sum_{\forall j}\left(\frac{\partial J}{\partial \mathbf{u}_j}\right)\bigg|_k (\mathbf{u}_j-\hat{\mathbf{u}}_j^k) \\
&0 \geq \mathbf{T}_p^k\left\{\mathbf{h}_p^k + \left(\frac{\partial \mathbf{h}_p}{\partial \mathbf{d}}\right)\bigg|_k (\mathbf{d}-\hat{\mathbf{d}}^k) + \left(\frac{\partial \mathbf{h}_p}{\partial \boldsymbol{\omega}}\right)\bigg|_k (\boldsymbol{\omega}-\hat{\boldsymbol{\omega}}^k) + \sum_{\forall j}\left(\frac{\partial \mathbf{h}_p}{\partial \mathbf{u}_j}\right)\bigg|_k (\mathbf{u}_j-\hat{\mathbf{u}}_j^k)\right\} \\
&0 \geq \mathbf{g}_p^k + \left(\frac{\partial \mathbf{g}_p^k}{\partial \mathbf{d}}\right)\bigg|_k (\mathbf{d}-\hat{\mathbf{d}}^k) + \left(\frac{\partial \mathbf{g}_p^k}{\partial \boldsymbol{\omega}}\right)\bigg|_k (\boldsymbol{\omega}-\hat{\boldsymbol{\omega}}^k) + \sum_{\forall j}\left(\frac{\partial \mathbf{g}_p^k}{\partial \mathbf{u}_j}\right)\bigg|_k (\mathbf{u}_j-\hat{\mathbf{u}}_j^k) \\
&0 \geq \mathbf{T}_p^k\left\{\mathbf{h}_q^k + \left(\frac{\partial \mathbf{h}_q}{\partial \mathbf{d}}\right)\bigg|_k (\mathbf{d}-\hat{\mathbf{d}}^k) + \left(\frac{\partial \mathbf{h}_q}{\partial \boldsymbol{\omega}}\right)\bigg|_k (\boldsymbol{\omega}-\hat{\boldsymbol{\omega}}^k)\right\} \\
&0 \geq \mathbf{g}_q^k + \left(\frac{\partial \mathbf{g}_q^k}{\partial \mathbf{d}}\right)\bigg|_k (\mathbf{d}-\hat{\mathbf{d}}^k) + \left(\frac{\partial \mathbf{g}_q^k}{\partial \boldsymbol{\omega}}\right)\bigg|_k (\boldsymbol{\omega}-\hat{\boldsymbol{\omega}}^k) \\
&0 = \boldsymbol{\omega} - \bar{\boldsymbol{\omega}}, \quad k \in K_{feas}
\end{aligned}
\tag{2.21}
$$

together with the integer cuts Eq. 2.8 and pure binary constraints of the original system. In Eq. 2.21, $\mathbf{T}_{\sqcup}^k$, $\sqcup = p, q$ are diagonal matrices with elements defined as:

$$
\mathbf{T}_{\sqcup,ii}^k = \begin{cases} -1 & \text{if } \lambda_{\sqcup,i}^k < 0 \\ 1 & \text{if } \lambda_{\sqcup,i}^k > 0 \\ 0 & \text{if } \lambda_{\sqcup,i}^k = 0 \end{cases}
\tag{2.22}
$$

where $\lambda_{\sqcup,i}^k$, $\sqcup = p, q$ are the Lagrange multipliers pertaining to the equality constraints $\mathbf{h}_p = 0$ and $\mathbf{h}_q = 0$. If control vector parameterization is used to solve the primal problem, then these multipliers and the gradients in Eq. 2.21 are available directly from the solution of the resulting NLP problem. If another type of dynamic optimization technique is used, e.g. multiple shooting, then the multipliers will still be directly available; however, the gradients have to be computed by performing an additional integration of the sensitivity or adjoint-based DAE system at the optimal solution of the primal problem [270].

Note that Eq. 2.21 bears some resemblance to the master problem proposed by Fletcher and Leyffer [271]. However, that work considers MINLP problems, not MIDO ones, where there are no equality constraints and where linearizations of inactive constraints are not included in the master problem.

2.9 New MIDO Algorithm II

A new algorithm, based on OA/ER, for the solution of the general MIDO problem Eq. 2.1 can now be summarized as follows:

Step 1. Set the termination tolerance, ϵ. Initialize the iteration counter, $k = 1$; lower bound, $LB = -\infty$; and upper bound, $UB = \infty$.

Step 2. For fixed values of the binary variables, $\boldsymbol{\omega} = \boldsymbol{\omega}^k$, solve the kth primal problem Eq. 2.3 to obtain a solution J^k. (For computational speed, it may be useful to omit the search variables and constraints that are superfluous due to the current choice of values for the binary variables.) Set $UB = \min(UB, J^k)$ and store the values of the continuous and binary variables corresponding to the best primal solution so far.

Step 3. Solve the primal problem Eq. 2.9 at the solution found in Step 2 with the full set of search variables and constraints included. Convergence will be achieved in one iteration. Obtain the gradients and Lagrange multipliers needed to construct the kth master problem Eq. 2.21.

Step 4. Solve the kth relaxed master problem Eq. 2.21, with integer cuts Eq. 2.8 and the pure binary constraints of the original system to obtain a solution η^k. Update the lower bound, $LB = \eta^k$.

Step 5. If $UB - LB \leq \epsilon$, or the master problem is infeasible, then stop. The optimal solution corresponds to the values stored in Step 2. Otherwise, set $k = k + 1$ and $\boldsymbol{\omega}^{k+1}$ equal to the integer solution of the kth master problem, and return to Step 2.

As with the GBD-based MIDO Algorithm I, if in Step 2 a particular set of values $\boldsymbol{\omega}^k$ renders the primal problem infeasible, then an infeasibility minimization problem can be solved, which in Step 3 may correspond to Eq. 2.13 [250]. The matrices $\mathbf{T}_p$ and $\mathbf{T}_q$ and gradients $\partial \mathbf{h}_p/\partial \mathbf{d}$, $\partial \mathbf{g}_p/\partial \mathbf{d}$, $\partial \mathbf{h}_q/\partial \mathbf{d}$, $\partial \mathbf{g}_q/\partial \mathbf{d}$ etc. are found from the solution of this problem and then used to add extra linearization constraints to the master problem Eq. 2.21.

2.10 Remarks on Algorithm II

(i) As already stated, unlike current MIDO approaches that use OA principles to construct the master problems, the algorithm outlined above is applicable to models where the binary variables appear explicitly in the DAE system. Like Algorithm I, Algorithm II has the advantage that it is independent of the type of method that is used for solving the dynamic optimization primal problem (although control vector parameterization may be preferred versus multiple shooting, for example, since then no intermediate problem needs to be solved to obtain the gradients that are necessary for constructing the master problem). Thus, again as with Algorithm I, Algorithm II allows existing dynamic optimizers to be used for solving MIDO problems. Algorithm II can also be readily extended to tackle problems involving time-dependent binary variables, in a similar fashion to that described in Sect. 2.7 for Algorithm I.

(ii) It should be noted that unlike the GBD-based MIDO Algorithm I, where the master problem Eq. 2.16 involves just one additional constraint per iteration, the master problem Eq. 2.21 in Algorithm II involves the addition of a large number of constraints at each iteration. These constraints correspond to linearizations about the optimal solution of the primal problem of the objective function, equality and inequality constraints. The MILP master problems in Algorithm II are therefore computationally more expensive than those in Algorithm I. (As noted in Sect. 2.2, however, for both GBD- and OA-based MIDO algorithms, the master problems are usually much less computationally demanding to solve than the primal, dynamic optimization problems.) Because the master problem in Algorithm II is more constrained than that in Algorithm I, the lower bound will, in general, be tighter, and so Algorithm II will often converge to a locally optimal solution of the MIDO problem in fewer iterations than Algorithm I. Note, however, that since most MIDO problems involve non-convexities, the additional constraints in the master problem of Algorithm II may cause it to cut off larger parts of the feasible region than Algorithm I. Thus, although Algorithm II may converge to a local optimum faster than Algorithm I, there is a chance that the solution will be worse.

2.11 Example—Binary Distillation Column

The application of the new algorithms for solving MIDO problems is now demonstrated with an example based on a binary distillation system from the MINOPT User's Guide [261]. The column has a fixed number of trays and the objective is to determine the optimal feed location (discrete decision), vapour boil-up, V, and reflux flow rate, R (continuous decisions), in order to minimize the integral square error (ISE) between the bottoms and distillate compositions and their respective set-points. The superstructure of the system is shown in Fig. 2.1. The following modelling assumptions were used by Schweiger et al.: (i) constant molar overflow; (ii) constant relative volatility, α; (iii) phase equilibrium; (iv) constant liquid hold-ups, equal to m for each tray and $10m$ for the reboiler and condenser; (v) no tray hydraulics; (vi) negligible vapour hold-ups; and (vii) no pressure drops. The system is initially at steady-state; at $t = 0$ there is a step change in the feed composition, z_f; and the inequality constraints are that the distillate composition must be greater than 0.98, and the bottoms composition must be less than 0.02, at the end of the time horizon of $400min$. The problem can be stated mathematically as:

Fig. 2.1 Superstructure for the binary distillation column example

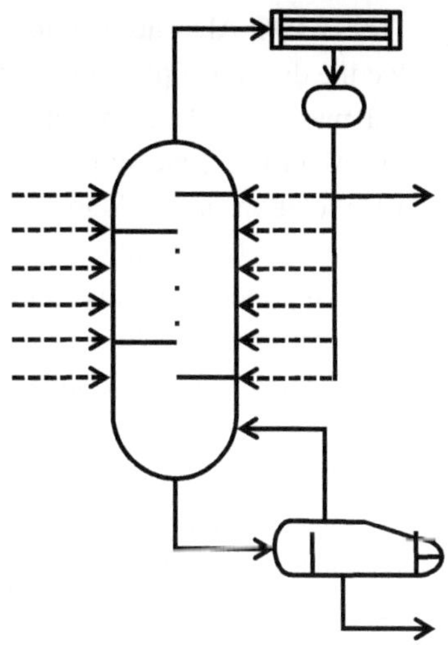

(1) Objective function

$$\min_{yf_1,\ldots,yf_N,V,R} \quad ISE(t_f)$$

$$\frac{d(ISE)}{dt} = (x_b - x_b^*)^2 + (x_{N+1} - x_{N+1}^*)^2$$

(2) Component balances

$$10m\frac{dx_b}{dt} = L_1x_1 - Vy_0 - Bx_b$$

$$m\frac{dx_i}{dt} = L_{i+1}x_{i+1} - L_ix_i + V(y_{i-1} - y_i) + Fyf_iz_f, \quad i = 1, \ldots, N-1$$

$$m\frac{dx_N}{dt} = -L_Nx_N + V(y_{N-1} - y_N) + Fy_Nz_f + Rx_{N+1}$$

$$10m\frac{x_{N+1}}{dt} = V(y_N - x_{N+1})$$

(3) Overall balances

(4) Vapour-liquid equilibrium

(5) Initial conditions

(6) Step disturbance

(7) End-point inequality constraints

(8) Time horizon

(9) Pure binary constraints

where yf_i is a binary variable that takes a value of 1 if tray i receives the feed, and is 0 otherwise. Only one tray is allowed to receive the feed, which is constrained to enter on tray 10 or above.

The MIDO Algorithms I and II were both implemented using gPROMS/gOPT v1.7 to solve the dynamic optimization primal problems, and GAMS/CPLEX v2.50 [272] for the MILP master problems. All the primal solutions, Lagrange multipliers and master solutions using Algorithm I are summarized in Table 2.1, while the primal solutions, gradients and master solutions using Algorithm II are summarized in Table 2.2. Using a termination tolerance of $\epsilon = 0$, both algorithms converged after four iterations, giving the same optimal solution reported by Schweiger et al. [261]. This corresponds to a feed location at tray 25, vapour boil-up of $1.5426 kmol/min$, reflux flow rate of 1.0024 kmol/min, and ISE of 0.1817. Note that integer cuts were added in each master problem to prevent the re-occurrence of previous solutions. For example, for the first master problem, the cut $(yf_{20} - \sum_{i \neq 20} yf_i \leq 0)$, was added.

It is interesting to observe that, for this example, the master problems of Algorithm II do not predict tighter lower bounds than the master problems of Algorithm I except on the fourth iteration when convergence of the overall MIDO problem has already been achieved. This behaviour can be understood in terms of the fact that, qualitatively, the difference between a GBD-based master problem and an OA-based one is that the former only considers the inequality constraints that are active at the primal solution, whereas the latter considers all the inequality constraints in the model [273]. Thus, if the inequality constraints that are inactive at the primal solution remain inactive at the solution of the OA master problem, then GBD and OA will give identical behaviour. This is demonstrated for this example in Appendix A.2.

2.12 Concluding Remarks

This chapter has presented an algorithm for solving MIDO problems that is based on generalized Benders decomposition principles with the use of a reformulation strategy. This algorithm has many advantages over current approaches, namely that (i) it is independent of the type of method used to solve the dynamic optimization primal problems; (ii) no intermediate adjoint problem needs to be solved in order to construct the master problems, even when the binary variables appear explicitly and non-linearly in the DAEs; and (iii) the master problems are significantly simpler. The algorithm can also be easily extended to handle the case of time-dependent binary variables, as shown in this chapter.

With the reformulation strategy used in Algorithm I, the chapter has also demonstrated how an algorithm can be developed that uses OA/ER principles to construct the master problems. This algorithm has similar properties to Algorithm I and thus, in contrast to current OA-based, MIDO algorithms, it is applicable to models where the binary variables appear explicitly in the DAEs.

Table 2.1 Progress of iterations of MIDO Algorithm I for the distillation column example [228]

Iteration number	1	2	3	4
Primal solutions				
Feed tray	20	26	24	25
V (kmol/min)	1.2231	1.7830	1.4111	1.5426
R (kmol/min)	0.6836	1.2426	0.8711	1.0024
ISE	0.1992	0.1827	0.1834	0.1817
UB	0.1992	0.1827	0.1827	0.1817
Lagrange multipliers				
ζ_1	0.4353	0.6147	0.6045	0.6204
ζ_2	0.3319	0.3965	0.4077	0.4087
ζ_3	0.2741	0.2864	0.3027	0.2984
ζ_4	0.2414	0.2307	0.2463	0.2406
ζ_5	0.2226	0.2024	0.2159	0.2101
ζ_6	0.2114	0.1881	0.1992	0.1939
ζ_7	0.2045	0.1808	0.1899	0.1853
ζ_8	0.2001	0.1771	0.1847	0.1806
ζ_9	0.1970	0.1753	0.1815	0.1780
ζ_{10}	0.1948	0.1745	0.1796	0.1765
ζ_{11}	0.1931	0.1742	0.1784	0.1757
ζ_{12}	0.1916	0.1742	0.1775	0.1752
ζ_{13}	0.1904	0.1744	0.1768	0.1749
ζ_{14}	0.1893	0.1746	0.1763	0.1747
ζ_{15}	0.1884	0.1749	0.1759	0.1746
ζ_{16}	0.1878	0.1752	0.1755	0.1745
ζ_{17}	0.1876	0.1755	0.1752	0.1745
ζ_{18}	0.1879	0.1759	0.1749	0.1745
ζ_{19}	0.1890	0.1763	0.1747	0.1745
ζ_{20}	0.1909	0.1766	0.1746	0.1746
ζ_{21}	0.1994	0.1770	0.1748	0.1747
ζ_{22}	0.2085	0.1773	0.1753	0.1750
ζ_{23}	0.2172	0.1777	0.1762	0.1754
ζ_{24}	0.2246	0.1779	0.1776	0.1759
ζ_{25}	0.2301	0.1777	0.1818	0.1763
ζ_{26}	0.2325	0.1765	0.1846	0.1763
ζ_{27}	0.2298	0.1698	0.1838	0.1721
ζ_{28}	0.2189	0.1537	0.1763	0.1600
ζ_{29}	0.1954	0.1204	0.1574	0.1339
ζ_{30}	0.1535	0.0566	0.1209	0.0850
Master solutions				
Feed tray	26	24	25	23
LB	0.1576	0.1813	0.1815	0.1848
$UB - LB \leq 0$?	No	No	No	Yes: STOP

Table 2.2 Progress of iterations of MIDO algorithm II for the distillation columns example [228]

Iteration number	1	2	3	4
Primal solutions				
Feed tray	20	26	24	25
V (kmol/min)	1.2231	1.7830	1.4111	1.5426
R (kmol/min)	0.6836	1.2426	0.8711	1.0024
ISE	0.1992	0.1827	0.1834	0.1817
UB	0.1992	0.1827	0.1827	0.1817
Objective gradients				
w.r.t. V, R	7.4445, −6.7611	12.108, −11.926	10.052, −9.7077	10.962, −10.701
yf_1, yf_2	−4.9048, −4.8013	−7.3116, −7.0977	−6.2937, −6.0972	−6.7519, −6.5413
yf_3, yf_4	−4.7434, −4.7107	−6.9898, −6.9352	−5.9924, −5.9361	−6.4315, −6.3740
yf_5, yf_6	−4.6918, −4.6806	−6.9075, −6.8934	−5.9057, −5.8891	−6.3438, −6.3277
yf_7, yf_8	−4.6737, −4.6692	−6.8863, −6.8827	−5.8798, −5.8746	−6.3191, −6.3144
yf_9, yf_{10}	−4.6662, −4.6639	−6.8809, −6.8802	−5.8715, −5.8695	−6.3118, −6.3103
yf_{11}, yf_{12}	−4.6622, −4.6608	−6.8799, −6.8798	−5.8683, −5.8674	−6.3095, −6.3090
yf_{13}, yf_{14}	−4.6596, −4.6587	−6.8800, −6.8802	−5.8668, −5.8663	−6.3087, −6.3085
yf_{15}, yf_{16}	−4.6581, −4.6580	−6.8805, −6.8808	−5.8659, −5.8656	−6.3084, −6.3084
yf_{17}, yf_{18}	−4.6588, −4.6611	−6.8812, −6.8817	−5.8655, −5.8655	−6.3085, −6.3087
yf_{19}, yf_{20}	−4.6657, −4.6743	−6.8823, −6.8831	−5.8660, −5.8673	−6.3092, −6.3101
yf_{21}, yf_{22}	−4.7129, −4.7640	−6.8845, −6.8869	−5.8700, −5.8754	−6.3119, −6.3153
yf_{23}, yf_{24}	−4.8258, −4.8989	−6.8913, −6.8998	−5.8856, −5.9049	−6.3220, −6.3349
yf_{25}, yf_{26}	−4.9855, −5.0899	−6.9163, −6.9487	−5.9830, −6.1007	−6.3594, −6.4534
yf_{27}, yf_{28}	−5.2187, −5.3816	−7.0640, −7.2642	−6.2709, −6.5180	−6.6042, −6.8391
yf_{29}, yf_{30}	−5.5935, −5.8772	−7.6064, −8.1968	−6.8835, −7.4364	−7.2083, −7.7990
Inequality values				
g_1, g_2	−3.9013e-3, 0	−2.2065e-3, 0	−2.9734e-3, 0	−2.6192e-3, 0
Gradients of g_1				
w.r.t. V, R	0.1819, −0.3292	0.0423, −0.0684	0.0821, −0.1403	0.0615, −0.1024
$yf_1, yf_2 (\times 1e6)$	−385.63, −415.24	−1.0724, −1.1254	−11.278, −12.010	−3.7995, −4.0215
$yf_3, yf_4 (\times 1e6)$	−433.12, −444.65	−1.1541, −1.1703	−12.433, −12.692	−4.1467, −4.2212
$yf_5, yf_6 (\times 1e6)$	−452.71, −458.81	−1.1796, −1.1838	−12.862, −12.976	−4.2676, −4.2958
$yf_7, yf_8 (\times 1e6)$	−463.70, −467.65	−1.1825, −1.1722	−13.045, −13.058	−4.3067, −4.2933
$yf_9, yf_{10} (\times 1e6)$	−470.56, −471.93	−1.1448, −1.0837	−12.981, −12.740	−4.2375, −4.1012
$yf_{11}, yf_{12} (\times 1e7)$	−4706.6, −4645.9	−9.5487, −6.9007	−121.89, −110.47	−38.088, −32.120
$yf_{13}, yf_{14} (\times 1e7)$	−4497.0, −4185.2	−1.5164, 9.3740	−87.721, −43.272	−20.205, 3.3290
$yf_{15}, yf_{16} (\times 1e6)$	−357.15, −239.88	3.1344, 7.5611	4.2762, 20.850	4.9574, 14.021
$yf_{17}, yf_{18} (\times 1e5)$	−1.9069, 39.354	1.6475, 3.4417	5.2700, 11.383	3.1759, 6.6453
$yf_{19}, yf_{20} (\times 1e5)$	116.14, 258.75	7.0530, 14.321	23.107, 45.587	13.429, 26.689
$yf_{21}, yf_{22} (\times 1e4)$	91.024, 181.41	2.8946, 5.8379	8.8680, 17.128	5.2608, 10.327
$yf_{23}, yf_{24} (\times 1e3)$	29.602, 43.746	1.1761, 2.3680	3.2961, 6.3307	2.0229, 3.9582
$yf_{25}, yf_{26} (\times 1e3)$	61.235, 83.232	4.7665, 9.5933	18.824, 38.274	7.7409, 22.509
$yf_{27}, yf_{28} (\times 1e2)$	11.157, 1.4905	2.7123, 5.8186	6.722, 11.031	4.6826, 8.5604
yf_{29}, yf_{30}	0.1998, 0.2700	0.1121, 0.2061	0.1754, 0.2751	0.1477, 0.2481
g_1, g_2	−3.9013e-3, 0	−2.2065e-3, 0	−2.9734e-3, 0	−2.6192e-3, 0

(continued)

Table 2.2 (continued)

Iteration number	1	2	3	4
Gradients of g_2				
w.r.t. V, R	−1.8800, 1.7074	−2.0744, 2.0432	−1.9993, 1.9309	−2.0286, 1.9803
yf_1, yf_2	1.1287, 1.1287	1.1473, 1.1481	1.1316, 1.1317	1.1347, 1.1349
yf_3, yf_4	1.1287, 1.1287	1.1484, 1.1486	1.1317, 1.1317	1.1350, 1.1351
yf_5, yf_6	1.1287, 1.12867 1.1487, 1.1488	1.1317, 1.1317	1.1350, 1.1351	
yf_7, yf_8	1.1286, 1.1286	1.1488, 1.1488	1.1317, 1.1317	1.1350, 1.1351
yf_9, yf_{10}	1.1286, 1.1286	1.1488, 1.1488	1.1317, 1.1317	1.1350, 1.1351
yf_{11}, yf_{12}	1.1286, 1.1286	1.1488, 1.1488	1.1317, 1.1317	1.1351, 1.1351
yf_{13}, yf_{14}	1.1287, 1.1287	1.1488, 1.1488	1.1317, 1.1317	1.1351, 1.1351
yf_{15}, yf_{16}	1.1288, 1.1289	1.1488, 1.1488	1.1317, 1.1318	1.1351, 1.1351
yf_{17}, yf_{18}	1.1292, 1.1296	1.1488, 1.1489	1.1318, 1.1319	1.1351, 1.1352
yf_{19}, yf_{20}	1.1305, 1.1322	1.1489, 1.1489	1.1320, 1.1323	1.1353, 1.1354
yf_{21}, yf_{22}	1.1398, 1.1504	1.1491, 1.1495	1.1328, 1.1338	1.1357, 1.1363
yf_{23}, yf_{24}	1.1639, 1.1804	1.1502, 1.1516	1.1356, 1.1392	1.1375, 1.1398
yf_{25}, yf_{26}	1.2009, 1.2267	1.1544, 1.1602	1.1539, 1.1767	1.1442, 1.1617
yf_{27}, yf_{28}	1.2599, 1.3038	1.1811, 1.2182	1.2107, 1.2614	1.1903, 1.2360
yf_{29}, yf_{30}	1.3632, 1.4455	1.2825, 1.3946	1.3378, 1.4551	1.3092, 1.4276
Master solutions				
Feed tray	26	24	25	23
V, R (kmol/min)	1.7488, 1.2072	1.5179, 0.9777	1.6375, 1.0977	1.3856, 0.8440
LB	0.1576	0.1813	0.1815	0.2093
$UB - LB \leq 0$?	No	No	No	Yes: STOP

The algorithms presented in this chapter are well-suited for the practical solution of large engineering problems using available tools. This offers the potential for much progress to be made in the solution of more realistic problems in areas such as simultaneous design, scheduling, and control problems as discussed in the following chapters.

3 PAROC: PARametric Optimisation and Control Framework

In this chapter, we present the main foundations and features of an integrated framework and software platform that enables the use of model-based tools in design, operational optimization and advanced control studies. A step-wise procedure is outlined involving: (i) the development of a high-fidelity dynamic model and its validation, (ii) a model approximation step including system identification, model reduction, and global sensitivity analysis, (iii) a receding horizon modeling step for model predictive control (MPC) and reactive scheduling, (iv) a suite of multi-parametric programming for optimization under uncertainty, explicit/multi-parametric MPC and state estimation, and (v) an in-silico validation step for the derived optimization, control and/or scheduling strategies to be analyzed and verified within the original high-fidelity model [227]. The software platform, PAROC, is also introduced briefly on a simple continuously stirred tank reactor example and demonstrated on a combined heat and power system in more depth.[1]

3.1 Overview

The integration of detailed modelling, design and operational optimization, controller design and scheduling/planning policies have been core process systems engineering challenges. While high-fidelity modelling and dynamic simulation have become standard engineering tools, with software systems such as ASPEN Plus and PSE's gPROMS becoming standard platforms,[2] the same is not true for the integration of design, control and scheduling. It is

[1] Reprinted from Chemical Engineering Science, vol. 136, E. N. Pistikopoulos, N. A. Diangelakis, R. Oberdieck, M. M. Papathanasiou, I. Nascu, and M. Sun, PAROC—An integrated framework and software platform for the optimisation and advanced model-based control of process systems, pp. 115–138, 2015, with permission from Elsevier [227].

[2] See http://www.aspentech.com/ products/aspen-plus.aspx and https://www.psenterprise.com/ products/gproms for more details.

B. Burnak et al., *Integrated Process Design and Operational Optimization via Multi-parametric Programming*, Synthesis Lectures on Engineering, Science, and Technology, https://doi.org/10.1007/978-3-032-10181-5_3

Table 3.1 An indicative list of model based control in the literature [227]

Authors	Remarks
Campo and Morari (1987) [274]	Min worst case ∞ norm
Lee and Yu (1997) [275]	Min worst case quadratic cost, use of dynamic programming for closed-loop
Schwarm and Nikolaou (1999) [276], Badgwell (1997) [277]	Min nominal objective s.t. robustness quadratic/linear constraints
Kassmann et al. (2000) [278]	apply robustness constraints to steady state target calculation
Scokaert and Mayne (1998) [279]	Min worst case quadratic, invariant set for stability
Lee and Cooley (2000) [280]	Min worst case quadratic cost s.t. quadratic constraints for stability
Mayne and Schroeder (1997) [281]	Min settling time, use invariant set
Bemporad et al. (2003) [282]	Min worst case $-\infty$ norm, use of dynamic programming, solve mpLPs consecutively

interesting to note that despite major advances in the areas of design and control, scheduling and control, and model-based advanced control (see Table 3.1), there is currently (i) no generally accepted methodology and/or protocol for such an integration and (ii) no commercially available software system to fully support such an activity. This can mainly be attributed to the following challenges: real-world systems are generally of high-complexity both mathematically and conceptually. Additionally, any software tackling such a problem would need to provide a unified approach not only towards modeling and model approximation, but also concerning solution of optimization problems as well as software integration. Lastly, any developed software platform needs to be tested and applied to problems of different complexities and structures which in return influences the design of the software. This requires a large interdisciplinary effort over an extensive period of time, posing an additional challenge for the development of such a system.

In this work, we present a comprehensive framework that enables the representation and solution of demanding model-based operational optimization and control problems following an integrated procedure featuring high-fidelity modeling, approximation techniques and optimization based strategies, including multi-parametric programming. A key advantage of the proposed framework is its ability to adapt to different classes of problems in an effortless manner through a prototype software platform PAROC (PARametric Optimisation and Control). The latter offers, among others, the great advantage of interoperability between advanced modeling software packages (PSE's gPROMS ModelBuilder), MATLAB based tools for model approximation and multi-parametric model-based controller and state estimator design tools.

The rest of the paper is structured as follows. Section 3.2 presents a brief overview of parametric programming before it describes a detailed description of the PAROC framework, the current state of its key components as well as the future outlook and perspectives. In Sect. 3.4, the PAROC framework is demonstrated on a residential combined heat and power system. Finally in Sect. 3.5, we present some concluding remarks on the presented software prototype.

3.2 The PAROC Framework

In this section we describe in detail the PAROC (PARametric Optimisation and Control) framework, which is depicted in Fig. 3.1.

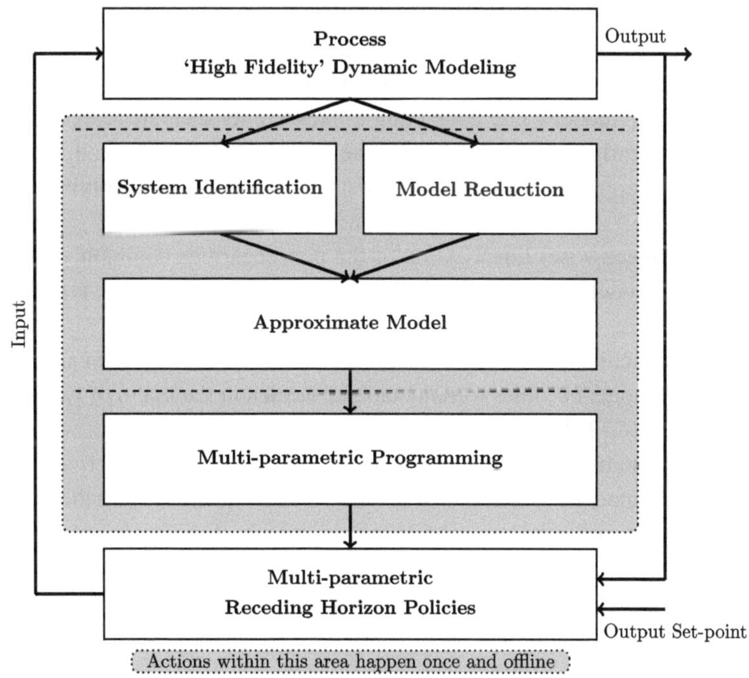

Fig. 3.1 The PAROC framework [Adapted from [227]]

3.2.1 High-Fidelity Modeling and Analysis

The first step of the PAROC framework is high-fidelity modeling and analysis. In particular, the scope is to (i) develop a high-fidelity model of the process [283, 284] based primarily on first principles, (ii) analyze the original problem e.g. using global sensitivity analysis [285–287] and (iii) perform parameter estimation and dynamic optimization of the developed model. Within our framework, the modeling software PSE's gPROMS ModelBuilder is used as it provides the aforementioned tools either directly or allows for their implementation via gO:MATLAB, a connection tool between MATLAB and gPROMS.

3.2.2 Model Approximation

Although it is possible to use a high-fidelity model for optimal design decisions, its complexity usually renders its direct use for the development of model-based strategies computationally expensive. The procedure of model approximation involves the reduction of complexity of the 'high-fidelity' model while attempting to preserve its accurate representation of the process at hand. Clearly, a trade-off between the two is of the essence in this procedure. A large variety of the approximation methods exists in open literature involving:

- model based (piece-wise) linearization of the non-linearities commonly present in the 'high-fidelity' model around one or multiple predefined points [288], typically coupled with
- model order reduction techniques, in cases where the state vector size of an already linear (or linearized) model becomes a computational burden in the realm of multi-parametric programming [289–291].
- data-driven methods where sets of input-output data are utilized to derive linear or non-linear surrogate models, standalone or in tandem with equations from the 'high-fidelity' model (black box and gray box models techniques are of the essence here [292, 293]) and
- commercially available algorithms that utilize one or various of the aforementioned characteristics (e.g. The System Identification Toolbox® of MATLAB®.).

In the PAROC framework, model approximation is addressed by the following two approaches:
System identification. A series of simulations of the high-fidelity model for different initial states is used to construct a meaningful linear state-space model of the process using statistical methods. One of the most widely applied tools within this area is the System Identification Toolbox from MATLAB.

Table 3.2 An indicative list of model approximation techniques in the context of multi-parametric model predictive control in the literature [227]

Authors	Remarks
Müller and Lemke (1995) [294]	Theoretical description of Group method of data handling (GMDH)
Sobol (2001) [295]	Description of Sobol's sensitivity analysis method as a variance based approach based on the anova decomposition. The sensitivity analysis indices are usually computed through Monte-Carlo numerical integration
Kontoravdi et al. (2005) [286], Kontoravdi et al. (2010) [296], Kiparissides et al. (2009) [287]	Use of sensitivity analysis in the context of biomedical engineering in order to asses the robustness of complex biological and biomedical models and quantify uncertainty
Homma and Saltelli (1996) [297], Saltelli et al. (2000) [285], Saltelli et al. (2010) [298]	Development of global sensitivity analysis as a tool to detect parameter interactions, in particular variance based methods
Rivotti et al. (2012) [299]	A model order reduction via empirical Grammians (Hahn and Edgar (2002) [300]) is combined with a mpMPC algorithm
Lambert et al. [289]	Using Monte-Carlo integrations, N step ahead affine representations are created

Model-reduction techniques. While system identification relies on the user in terms of interpretation of the data and processing of the results, model-reduction techniques somewhat "automate" the reduction process based on formal techniques (see Table 3.2 for recent contributions).

3.2.3 Multi-parametric Programming

After the model approximation step, a state-space model is obtained which is used for the development of receding horizon policies. The calculation of such policies, e.g. in the form of control laws or scheduling policies, traditionally requires the online solution of an optimization problem, which might be computationally intractable [301]. Therefore, the PAROC framework employs multi-parametric programming, where the optimization problem is solved offline as a function of a set of bounded parameters. In addition, depending on the cost function and the characteristic of the system considered, the complexity of the optimization problem changes considerably. A review on key development in multi-parametric programming is given in Table 3.3, while Table 3.4 lists the resulting multi-parametric programming problems for the most common cost functions ($1/\infty$-norm, 2-

Table 3.3 An indicative list of multi-parametric programming in literature [227]

mpP	Authors	Theoretical development
mpLP	Gal and Nedoma (1972) [302], Gal and Davis (1978) [303]	Linear objective and constraints: extension of simplex algorithm—basis exchange
	Bemporad et al. (2000) [304], Dua et al. (2002) [305]	Solve Linear Program (LP) and use sensitivity analysis to derive expressions. Use inactive inequalities and Lagrangian multipliers to define regions. Degenerate solutions are avoided
mpNLP	Dua et al. (1999) [306], Dua et al. (2004) [307]	Non-convex optimization. Generate convex overestimator and underestimator. Linearization of convex functions. mpNLP on the functions. Employ a spatial branch-and-bound to obtain solution
	Dua et al. (2002) [305]	Quadratic objective and linear constraints (mpQP). Uses sensitivity analysis from Fiacco and Kyparisis (1986) [308] to obtain the exact expressions under convexity assumptions
	Zafiriou (1990) [309]	Quadratic programs. Obtain solutions for different sets of active constraints
	Fiacco and Kyparisis (1986) [308], Fiacco and Ishizuka [310]	Convex problem. Sensitivity analysis around optimal point. Linearization to obtain expressions
	Benson (1982) [311]	Non-convex problem. Convex over- and underestimators

norm) and system characteristics (continuous and hybrid systems). Thus, PAROC relies on efficient algorithms and software tools for the solution of these different classes of multi-parametric programming problems.

The rest of the section first outlines the current state-of-the-art in different fields of multi-parametric programming, before key recent advances are discussed and an outlook on future research directions is presented.

Table 3.4 Different classes of multi-parametric programming (mpP) problems [227]

Cost function system class	mpP	Problem formulation	Key references
$1/\infty$-norm Cont. system	mpLP	$z(\theta) = \min_x \ (H\theta + c)^T x$ $s.t. \ Ax \le b + F\theta$ $x \in \mathbb{R}^n$	Gal and Nedoma (1972) [302], Bemporad et al. (2002) [312], Borrelli et al. (2005) [313], and Khalilpour and Karimi (2014) [314]
2-norm Cont. system	mpQP	$z(\theta) = \min_x \ (Qx + H\theta + c)^T x$ $s.t. \ Ax \le b + F\theta$ $x \in \mathbb{R}^n$	Bemporad et al. (2002) [83], Dua et al. (2002) [305], Tøndel et al. (2003) [315], Mayne and Raković (2003) [316], Spjøtvold et al. (2006) [317], and Feller and Johansen (2013) [318]
$1/\infty$-norm Hybrid system	mpMILP	$z(\theta) = \min_\omega \ (H\theta + c)^T \omega$ $s.t. \ Ax + Ey \le b + F\theta$ $x \in \mathbb{R}^n, y \in \{0, 1\}^p$ $\omega = [x^T \ y^T]^T$	Acevedo and Pistikopoulos (1997) [319], Li and Ierapetritou (2007) [320], Dua and Pistikopoulos (2000) [321], Wittmann-Hohlbein and Pistikopoulos (2012) [322], Rivotti and Pistikopoulos (2014) [323], Oberdieck et al. (2014) [324] and Charitopoulos et al. [325]
2-norm Hybrid system	mpMIQP	$z(\theta) = \min_\omega \ (Q\omega + H\theta + c)^T \omega$ $s.t. \ Ax + Ey \le b + F\theta$ $x \in \mathbb{R}^n, y \in \{0, 1\}^p$ $\omega = [x^T \ y^T]^T$	Dua et al. (2002) [305], Almér and Morari (2013) [326], Axehill et al. (2014) [327], Oberdieck et al. (2014) [324], and Han and Chen (2015) [328]

3.2.3.1 Current State

Multi-parametric linear programming (mpLP). The two currently available mpLP algorithms are distinguished by their significantly different approaches to the solution. The first algorithm finds the optimal basis by evaluating a connected graph [302]. The second algorithm explores the parameter space using the properties of the generated polytopes [305, 312]. In addition, a recent contribution investigated the case of a parameter dependence in the constraint matrix [314].

Multi-parametric linear programming (mpQP). The two main approaches for solving mpQP problems are (i) a geometrical approach as presented for mpLP problems [305, 312, 315, 317] and (ii) a combinatorial (also called "reverse transformation") approach [316, 318, 329], where the combination of all possible active sets of constraints is considered.

Multi-parametric mixed-integer programming (mpMILP) problem. In this area, two major strategies for the solution of mpMILP problems have been outlined: (i) a branch-and-bound approach in conjunction with suitable comparison and fathoming procedures [319]

Table 3.5 An indicative list of multi-parametric mixed-integer programming algorithms in the literature [227]

mpMIP	Authors	Theoretical development
mpMILP	Acevedo and Pistikopoulos (1997) [319], Acevedo and Pistikopoulos (1999) [330], and Oberdieck et al. (2014) [324]	Modified branch and bound. Comparison procedure between parametric solutions
	Dua and Pistikopoulos (2000) [321]	Decomposition into (m)pLP and MILP
	Crema (2002) [331]	mpILP: iterate between an MILP free θ and ILP with fixed θ. Stop when no improved realization is found
mpMINLP	Dua et al. (1999) [306]	Convex systems. Iterate between a mpNLP and a MINLP
	Acevedo and Pistikopoulos (1996) [330], Pertsinidis et al. (1998) [332], Papalexandri and Dimkou (1998) [333], and Dua et al. (1999) [306]	Convex systems. Iterate between a (m)pNLP and a (m)pMILP

and (ii) a decomposition-type approach which alternates between a mixed-integer linear programming (MILP) and a mpLP problem [321].

Multi-parametric mixed-integer quadratic programming (mpMIQP) problem. Similar to mpMILP problems, a decomposition-type approach can be used for mpMIQP problems. This type of approach relies on global optimization as shown by Dua et al. (2002) [305]. In addition, strategies based on a branch-and-bound approach have been published recently [324, 327]. Furthermore the solution space of mpMIQP problems has been described and quasi-Lipschitz continuity of the solution has been proven [328].

A review on key advances in multi-parametric mixed-integer programming is shown in Table 3.5.

The interested reader can refer to [334] for a full review on multi-parametric programming.

3.2.3.2 Noteworthy Developments

mpMILP and mpMIQP problems. For mpMILP problems, there have been major advances in two areas: (i) an algorithm for the solution of mpMILP problems with parameter dependence in the constraint matrix has been presented [322, 335, 336] and (ii) a dynamic programming approach for hybrid control problems which result in a mpMILP was shown

[323, 337]. In the area of mpMIQP problems, the recent contributions [327] have opened up the area for a more generalised approach to this topic. In particular, the similar structures between the algorithms presented provide a framework for the solution of mpMIQP problems.

Multi-parametric Dynamic Optimization (mpDO) problems. For mpDO problems, where the dynamic process is represented by differential equations, the approaches used to obtain optimal solution can be divided into two main routines. One approach is by transforming the optimization problem into a finite dimensional multi-parametric programming problem by control vector parameterization [253, 270] and integration of the dynamic system [338–340], thus enabling the use of general algorithms of multi-parametric programming to generate optimal control laws as an explicit function of varying parameters [341]. The other method, designed for constrained linear quadratic optimal control problem, is based on calculus of variations by implementing the minimum principle [342]. With certain constraints active, the optimal conditions for Hamiltonian systems corresponding to continuous time optimal control problem can be expressed as certain boundary value problem [343–345]. Complete mapping from system parameters to control variables can be obtained in continuous time form by solving a set of differential and algebraic equations of the boundary value problems [346, 347].

A table with a literature overview over parametric programming and sensitivity analysis in dynamic optimization is shown in Table 3.6.

3.2.4 Multi-parametric Moving Horizon Policies

While multi-parametric programming has been applied in a variety of areas, a key application lies in the offline calculation of moving horizon policies such as control laws and scheduling policies. The underlying idea is to consider the states of the system as parameters, and thus solve the optimization problem over a range of admissible states.

Remark 1 In addition, measured disturbances, if present, are also considered as parameters as well as state-space and high-fidelity model mismatch and the output setpoint.

In general, we consider the following optimisation problem:

Table 3.6 an indicative list of multi-parametric programming and dynamic optimization [227]

Authors	Theoretical development
Solís-Daun et al. (1999) [348]	Stability of a dynamic system subject to uncertainty. Optimal feedback control as a function of parameters. No constraints included. Cannot be used for range of variations
Bemporad and Morari (1999) [349]	A framework for modeling and controlling systems described by interdependent physical laws, logic rules, and operating constraints, denoted as mixed logical dynamical (MLD) systems
Dontchev et al. (2000) [350]	Optimal control. Solution stability under perturbations. Expressions for neighbourhood around the optimal solution. Does not consider range of variations
Malanowski and Maurer (2001) [344], Augustin and Maurer (2001) [345]	Sensitivity analysis of non-linear optimal control problems. Compute derivatives of optimal conditions as a function of parameters. Problems with high order path constraints
Diehl et al. (2002) [351]	On-line Dynamic Optimization. Use derivatives of objective and constraints with respect to parameters to derive a perturbation manifold. Use it as an approximation to the dependence of the optimal conditions. Does not obtain complete profile of optimal conditions in terms of parameters
Bansal et al. (2003) [228]	New formulations and algorithms for solving mixed-integer dynamic optimization (MIDO) problems. The algorithms are based on decomposition into primal, dynamic optimization and master, mixed-integer linear programming sub-problems
Sakizlis et al. (2005) [346]	An algorithmic framework is presented for the derivation of the explicit optimal control policy for continuous-time linear dynamic systems that involve constraints on the process inputs and outputs. The control actions are usually computed by regularly solving an on-line optimization problem in the discrete-time space based on a set of measurements that specify the current process state
Faísca et al. (2007) [352]	A global optimization approach for the solution of various classes of bilevel programming problems (BLPP) based on recently developed parametric programming algorithms

$$\begin{aligned} V_N^*(x_0) &= \min_{U \in \mathcal{U}} \quad J(U, X) \\ &= \min_{U \in \mathcal{U}} \quad \|x_N\|_P^p + \sum_{k=0}^{N-1} \|x_k\|_S^p + \|u_k\|_R^p \\ s.t. \quad & x_{k+1} = Ax_k + Bu_k + Cd_k \\ & y_k = Dx_k + Eu_k + Fd_k + e \\ & h(u_k, x_k, y_k) \leq 0 \\ & x_{k+N} \in \mathcal{X}_T \end{aligned} \tag{3.1}$$

where u, x, and y are the moving horizon policies, states and outputs of the considered system, respectively, $U = [u_0, ..., u_{N-1}]$, $\|\cdot\|^p$ is the p-norm and P, S, and R are the corresponding weights. In addition, x_0 are the initial states and the set $\mathcal{X}_T$ is the terminal set that, if well-defined, ensures stability [190].

Remark 2 The parameter dependence of the objective function can be avoided using the Z-transformation [83], i.e.

$$z = u + H^{-1}F^T x, \tag{3.2}$$

where H is the Hessian of the objective and F is the bilinear term matrix between x and u.

While Problem 3.1 describes the general case of moving horizon policies, the remaining part of the section will focus predominately on multi-parametric model-predictive control (mpMPC). This is due to the extensive number of contributions and advances that have been made in this field.

3.2.4.1 Current State

Nominal mpMPC. For the case of nominal mpMPC problems, numerous publications have appeared (see Bemporad et al. (2002) [83], Pistikopoulos (2009) [301]). Although from a theoretical aspect little advances have been made, nominal mpMPC has been applied in a variety of fields such as electric drives [353], type 1 diabetes [354], and miniature helicopters [318].

Hybrid mpMPC. In general, hybrid mpMPC refers to the multi-parametric model-based control of systems that are described by both linear and logical dynamics [349, 355]. While different algorithms are available to solve hybrid mpMPC problems [305, 313, 324, 327], their applicability is still limited due to the high computational burden associated with the solution of multi-parametric mixed-integer programming (mpMIP) problems.

Robust mpMPC. Robustness describes the immunity of a system against perturbations [356]. In mpMPC, efforts have been made to develop a multi-parametric model-based controller that guarantees performance while satisfying the constraints, termed robust mpMPC. While early approaches focused on (i) additive disturbances [357–359] (ii) model uncertainties [282, 360], (iii) min–max robust mpMPC [282, 358, 361] and (iv) linear input/output models [359], recent advances point towards a more general approach for robust mpMPC [360, 362]. The key idea is (i) a dynamic programming reformulation of the original problem, (ii) the formulation of the robust counterpart and (iii) the solution of the resulting mpLP problem.

3.2.4.2 Recent Developments

Integration of scheduling and control. Similar to MPC, multi-parametric programming approaches for scheduling have been presented in the past [320, 363–365]. In addition, recently developed representation of scheduling problems in a state-space model [128] has allowed for the applicability of the concepts of mpMPC in scheduling [152] (see also Chapter 5).

Development of mpMPC for periodic systems. Applying MPC to periodic systems is a major challenge due to the presence of numerical computation and robustness issues [366]. In the realm of mpMPC, studies on pressure-swing adsorption [283, 366] have shown the applicability of mpMPC to such systems, however a comprehensive theory for general periodic systems also involving strong nonlinearities is still lacking.

Use of mpMPC in biomedical applications. While mpMPC has found applications within biomedical sciences [354], the general complexity of biological and biomedical systems makes a model-based optimization approach rather challenging. Applications of PAROC to such systems include leukemia [367] and anesthesia [368].

3.2.5 Multi-parametric Moving Horizon Estimation

Moving horizon estimation (MHE) is an estimation method based on optimization that considers a limited amount of past data. One of the main advantages of moving horizon estimation is the possibility to incorporate system knowledge as constraints in the estimation. In MHE the system states and disturbances are derived by solving the following optimisation problem [190, 369, 370]:

$$\begin{aligned}
\min_{\hat{x}_{T-N/T},\hat{W}_T} \quad & \|\hat{x}_{T-N/T} - \underline{x}_{T-N/T}\|_{P^{-1}_{T-N/T-1}} - \|Y^{T-1}_{T-N} - O\hat{x}_{T-N/T} - \bar{c}bU^{T-2}_{T-N}\|^2_{P-1} \\
& + \sum_{k=T-N}^{T-1} + \|\hat{w}_k\|_{Q_k^{-1}} + \|\hat{v}_k\|_{R_k^{-1}} \\
s.t. \quad & \hat{x}_{k+1} = A\hat{x}_k + Bu_k + G\hat{w}_k, \quad y_k = C\hat{x}_k + \hat{v}_k \\
& \hat{x}_k \in X, \hat{w}_k \in \Theta, \hat{v}_k \in V
\end{aligned} \tag{3.3}$$

where T is the current time step, $Q_k \succ 0$, $R_k \succ 0$, $P_{T-N/T-1} \succ 0$, are the covariances of w_k, v_k, x_{T-N}, respectively and assumed to be symmetric, N is the horizon length of the MHE, Y^{T-1}_{T-N} is a vector containing the past $N+1$ measurements and U^{T-1}_{T-N} is a vector containing the past N inputs. x, v, and w denote the variables of the system and $\hat{x}$, $\hat{v}$, and $\hat{w}$ denote the estimated variables of the system and $\hat{x}_{T-T/N}$ and $\hat{W}_T = W^{T-1}_{T-N}$ denote the decision variable of the optimization problem, the estimated state variable and the noise, respectively.

To derive the multi-parametric counterpart (mpMHE) formulation, the problem in Eq. 3.3 is reformulated as a multi-parametric programming problem.

$$\begin{aligned}
\min_{\hat{x}_{T-N/T},\hat{W}^{T-1}_{TN/T}} \quad & \frac{1}{2}\|[\|]\left[\hat{x}_{T-N/T} \ \hat{W}^{T-1}_{TN/T}\right]^2_H + \theta f\left[\hat{x}_{T-N/T} \ \hat{W}^{T-1}_{TN/T}\right] \\
s.t. \quad & K\left[\hat{x}_{T-N/T} \ \hat{W}^{T-1}_{TN/T}\right] \leq k
\end{aligned} \tag{3.4}$$

The parameters of the multi-parametric programming problem Eq. 3.4 are the past and current measurements and inputs and the initial guess for the estimated states.

3.2.5.1 Current State

Unconstrained MHE. There are a few necessary steps that lead to incorporating the constrained MHE into robust MPC. The estimation error at the beginning of the horizon and at the current time have to be derived and the bounding set of the estimation error has to be obtained. The unconstrained moving horizon estimator is equivalent to the Kalman filter [316], the estimation error and the bounding sets they generate should be equivalent so the Kalman filter can be used for comparison.

Constrained MHE. In order to formulate and solve the constrained moving horizon estimator with multi-parametric programming, the optimization problem is reformulated into the standard multi-parametric quadratic form. Previous work has been performed on reformulating the MHE with the filtered arrival cost [371] and with the smoothed update of the arrival cost [372].

3.2.5.2 Noteworthy Developments

MHE with smoothed arrival cost. The formulation of the MHE with the smoothed arrival cost is still an open issue in the literature. The optimization problem is reformulated into the standard multi-parametric quadratic form. The smoothed update of the arrival cost involves fewer parameters in the multi-parametric formulation of the MHE and hence it requires less computational effort to solve the mpMHE than the equivalent estimation problem that utilises the filtered arrival cost [373].

Simultaneous mpRHE and mpMPC. The implementation of explicit/multi-parametric MPC, and in general, MPC, is based on the assumption that the state values are readily available from the system measurements and also that the measurable output is free of noise influence. However in reality, the measured output may be noisy and the system measurements do not offer this information directly. Instead, the state information needs to be inferred from the available output measurements by using a state estimator which obtains an estimate $\bar{x}$ of the real state x. The framework uses the constrained MHE that gives improved estimation results compared to the unconstrained estimators. The estimation error remains inside the calculated error set and hence the MPC guarantees to satisfy the constraints [374].

Simultaneous mpRHE and mpMPC for biomedical applications. Biomedical systems are typically modelled as complex systems with a high degree of nonlinearity. Estimation techniques play an important role in such processes since some of the parameters and the states of the systems cannot always be measured directly from the system outputs. In most of the biomedical applications, the optimal policies rely on patient-dependent data which might be unavailable or computationally impossible to retrieve in a reasonable time frame. This makes simultaneous mpMHE and mpMPC an important ongoing research area that can deal with some critical issues especially on topics such as intravenous and volatile anaesthesia, type-1 diabetes and leukemia.

3.2.6 Remarks

In the following several comments concerning multi-parametric programming and its application are made:

Problem size. Albeit avoiding the online solution of optimization problems, deriving the solution of multi-parametric programming problems is challenging. In particular, the added complexity from a software point of view often provides a limit on certain classes of problems which can be solved. However, with increasing computing power as well as theoretical advances such as parallelization, the number of problems which can be solved is ever increasing.

Problem selection. Albeit its advantages, explicit control and scheduling is not the only viable strategy in this area. For example in many cases, the use of classical MPC is sufficient in order to provide an optimal control strategy, e.g. if the feedback time is significantly larger than the time required to solve the corresponding optimisation problem online. Thus, multi-parametric programming is not considered as a tool to replace other approaches, but to allow for the solution of very challenging control problems such as very complex systems (periodic systems, hybrid systems etc.) where the time needed to solve the optimization problem is too large with respect to the feedback time. Another important application lies in the field of biomedical engineering, as proven by the successful real-world application of explicit control for Diabetes-I patients [375]. In the later, the a priori knowledge of the full map of possible solutions play a vital role.

3.2.7 Software Implementation and Closed-Loop Validation

3.2.7.1 Multi-Parametric Programming Software

In conjunction with the aforementioned theoretical developments, PAROC provides software solutions to key aspects of the framework.[3] In particular, it offers tools for the formulation and solution of multi-parametric programming problems. Based on POP [376], it contains the state-of-the-art algorithms which allow for an efficient solution of mpLP, mpQP, mpMILP and mpMIQP problems. Furthermore, its interconnection with gPROMS ModelBuilder makes the use of the PAROC framework straightforward and allows for an intuitive approach for design, operation and control problems.

3.2.7.2 Integration of PAROC in gPROMS ModelBuilder

The developed multi-parametric moving horizon policies and estimators are validated in a closed-loop fashion against the original high-fidelity model. However, within the PAROC framework, the high-fidelity modeling and analysis is performed in gPROMS ModelBuilder, while the model reduction as well as the formulation and solution of the multi-parametric programming problem is carried out in MATLAB. Thus currently, the closed-loop validation of the developed controller is done in the MATLAB environment using the gPROMS ModelBuilder tool gO:MATLAB. While this is a valid way of performing closed-loop validation, this does not allow for the use of the tools available in gPROMS (e.g. dynamic optimization, sensitivity analysis, experiment design). In Fig. 3.2 a schematic that depicts the interaction between the different software packages is demonstrated. The actions enclosed in the grey rectangle happen in the MATLAB environment. While the procedure in extracting the high-fidelity model from gPROMS to a MATLAB friendly format is simple, straightforward and thoroughly described in the user guides, the design of the controller in the POP software [376] as well as interfacing the multiparametric solution to the modeling software

[3] http://paroc.tamu.edu/.

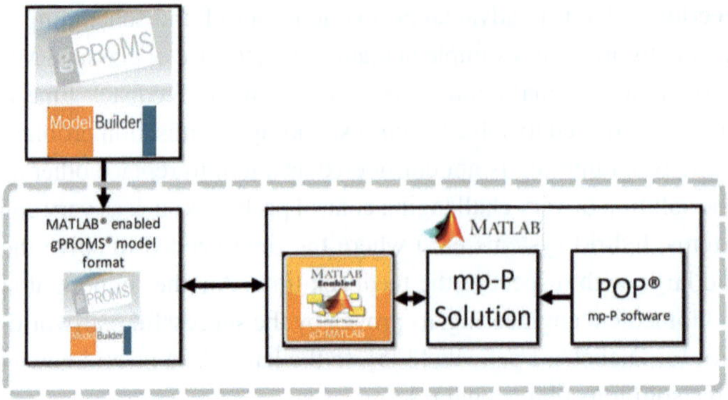

Fig. 3.2 Closed-loop validation step in the PAROC framework through gO:MATLAB [227]

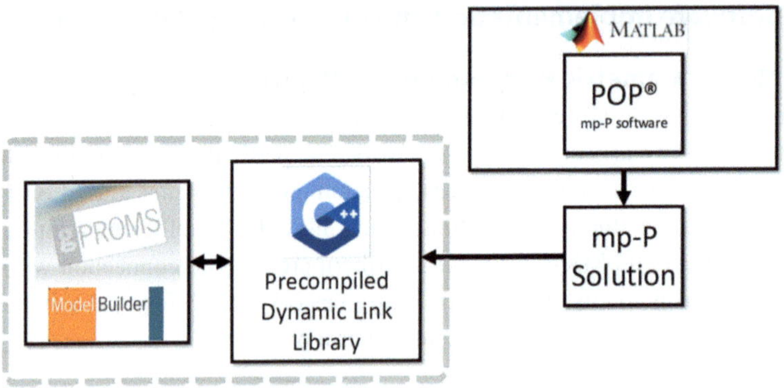

Fig. 3.3 Closed-loop validation step in the PAROC framework in the gPROMS environment through C++ [227]

via gO:MATLAB requires a good knowledge of both software packages. More specifically, the user needs to specify the information that is exchanged (including uncertain parameters, input actions, outputs and output setpoints) in both ends.

Therefore, we have developed a software solution that enables the direct export of the mpMPC controller developed in MATLAB into gPROMS ModelBuilder as a foreign object. This foreign object, coded in C++, loads the matrix representation and provides a simple look-up table as part of the gPROMS ModelBuilder architecture, similar to a Proportional-Integral-Derivative (PID) controller. Similar to the gO:MATLAB case, a schematic is provided (see Fig. 3.3). In this case, the actions within the grey rectangle happen within the gPROMS platform.

3.3 Illustrative Example

In this section, we demonstrate a step-by-step walk-through of the PAROC framework to develop an explicit MPC on a simple continuously stirred tank reactor. The aim of the controller is to track the concentration (C_P) set points which change arbitrarily within certain parameter bounds by manipulating the reactant volumetric flow rate Q. The stoichiometry of the reaction is $2R \rightarrow P$.

High fidelity model. We start with developing an accurate representation of the system dynamics. One can utilize first principle models, data-driven models, or a combination of the two classes of techniques. The process of interest in this case study is described by the following set of diferential algebraic equations (DAE). First, we have the mass balance equations around the CSTR, as presented by Eq. 3.5.

$$\begin{aligned} \frac{dR}{dt} &= \frac{Q}{V}(R_f - R) - 2r \\ \frac{dP}{dt} &= \frac{Q}{V}P + r \end{aligned} \tag{3.5}$$

where R is the reactant concentration, V is the volume of the CSTR and equals to 0.6 m^3, r is the rate of reaction, and R_f denotes the reactant concentration at the feed and equals to 1 mol/m^3. The reaction is assumed to be first order, therefore the power law reaction rate is formulated as follows.

$$r = kR^2 \tag{3.6}$$

where k is the rate constant and equals to 0.1 kmol/m^3h. The reaction stoichimetry and the parameters of this example can be found from Flores-Tlacuahuac and Grossmann (2006) [90].

Model approximation. Although the CSTR model given by Eqs. 3.5 and 3.6 is simple, it is nonlinear due to the quadratic power law kinetic expression. Therefore, we cannot derive a simple linear MPC for this problem. Therefore, we appeal to the System Identification Toolbox of MATLAB to develop an affine state space representation of the model that has satisfactory accuracy within the range of operation. We generate an input profile using pseudorandom binary sequence (PRBS) and feed to the model to observe the process outputs. The developed state space matrices are as follows.

$$A = \begin{bmatrix} 0.86 & -0.01 & 0.01 & -0.01 & 0.07 & 0.03 \\ -0.05 & 1.00 & -0.00 & -0.07 & 0.03 & 0.04 \\ 0.06 & 0.01 & 1.00 & 0.17 & 0.11 & -0.09 \\ 0.01 & 0.01 & -0.16 & 0.69 & 0.65 & 0.25 \\ 0.00 & 0.00 & -0.00 & -0.02 & 0.60 & -0.68 \\ 0.00 & -0.00 & -0.00 & -0.00 & -0.03 & 0.90 \end{bmatrix}, B \times 10^3 = \begin{bmatrix} 2.3 \\ 1.2 \\ 1.7 \\ 4.2 \\ 0.4 \\ 0.1 \end{bmatrix} \quad (3.7)$$
$$D = \begin{bmatrix} 2.91 & -9.86 & -1.68 & 0.42 & -0.09 & 0.04 \end{bmatrix}$$

The time step of the approximate model is 1 min.

Multi-parametric Model Predictive Control. After developing an affine state space representation for the CSTR, we use Eq. 3.1 to develop the explicit optimal control law. We use the POP toolbox [376], which is available in the MATLAB environment.

Closed loop validation. The developed explicit control law is embedded in the high fidelity model, and the closed loop simulation is obtained without solving a single optimization problem online. Figure 3.4 shows the closed loop behavior of the system with the developed control law. Here, the controller first brings the system to the first set point and then transitions to the second set point with sufficient accuracy. Therefore, the developed explicit control law is validated against the high fidelity model.

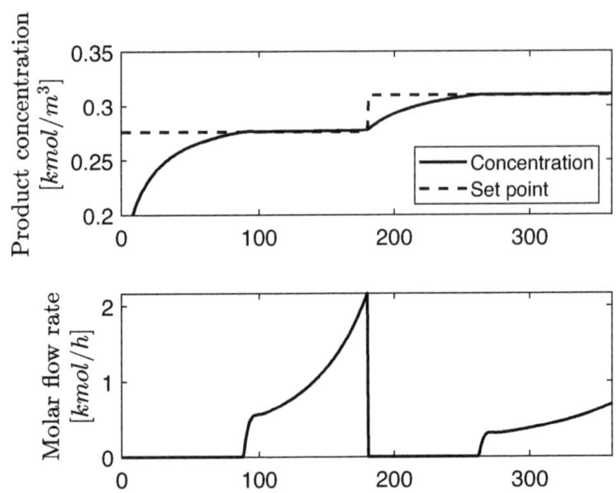

Fig. 3.4 Closed loop simulation for the CSTR example

3.4 A Case Study: Residential Combined Heat and Power System

This section presents the application of the PAROC framework for the design of the control scheme and the development of multi-parametric model predictive controllers for a residential scale cogeneration system. Typically, the use of cogeneration systems aims to reduce the system emissions and increase its operational efficiency via the simultaneous production of usable heat and electrical power in a single process, consuming the same amount of fuel [377]. Substantial work has been performed over the last years in the field of cogeneration, employing systems of various scales and power production technologies, focusing on both operational and design aspects [377–380]. Evidently, the rising interest of such systems requires further attention.

Here, we present an internal combustion powered cogeneration system for residential use that utilises natural gas as a fuel. The high fidelity model of the CHP system features a detailed description of the internal combustion engine based on a mean value approach and sub-models for the throttle valve, the intake and exhaust manifolds, and the external circuit, based on first principles. A detailed representation of the system and its components is presented in Fig. 3.5. The system is divided into two subsystems, namely the power generation subsystem and the heat recovery subsystem. In this decentralized approach, the two subsystems are treated separately in terms of controller development. The objective of the system is to produce water of certain temperature and flow, i.e. the water temperature and mass flow rate of the heat recovery subsystem are treated as control variables, while the mass flow rate setpoint and the operational level of the power generation subsystem are treated as manipulated variables. The latter is treated as a power output setpoint for the power generation subsystem. In the following sections, a step-by-step implementation of the PAROC Framework is demonstrated through this CHP case study.

High-fidelity modeling. The CHP system of Fig. 3.5 consists of a large number of components which have been modelled independently. The interactions of the components manage to adequately describe the operation of the CHP system [377]. Table 3.7 categorises those components based on their respective subsystem.

The throttle valve manipulates (i) the air inlet mass flow rate, directly and (ii) the injected fuel inlet mass flow rate, indirectly. It has been modelled as a typical pneumatic butterfly valve. The intake manifold distributes the inlet air into the engine cylinders, while the exhaust manifold gathers the post-combustion, high temperature exhaust gases. The main contribution of the manifold models is to manipulate the pressure of the fluid streams. Typical model principles of a small tank have been used while the shape and the size of the manifolds are subject to the engine specifications. It is assumed that the pressure at the outlet of the intake manifold is equal to atmospheric and the pressure at the outlet of the exhaust manifold exceeds the atmospheric by 20%.

The reciprocating internal combustion engine is the system's prime mover, responsible for the generation of kinetic power. Based on the modeling principles described by Guzzella

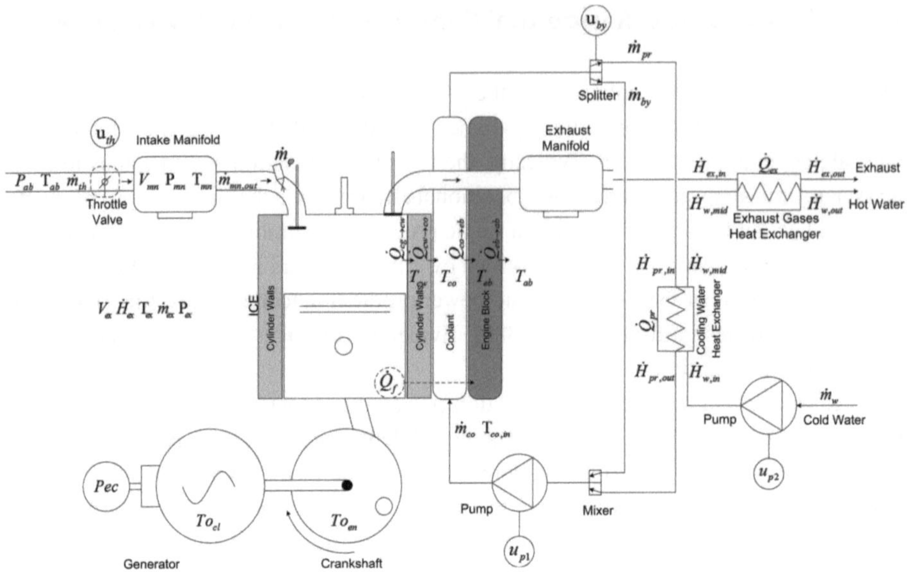

Fig. 3.5 Flowsheet of an internal combustion engine powered residential CHP system [227]

Table 3.7 System components of the CHP per subsystem [227]

Subsystem	Component
Power generation	Throttle valve
	Intake & exhaust manifolds
	Internal combustion engine
	Electrical generator
Heat recovery	Engine inner cooling system
	Heat exchangers
	Outer cooling system

and Onder (2010) [381] and Heywood (1989) [382] a mean-value model was developed to describe the engine operation. The model accounts for both the torque generation and the heat generation as a byproduct of the natural gas combustion. The electrical generator is the last part of the power generation subsystem, responsible for transforming the engine torque, via the crankshaft into electrical power.

The heat recovery subsystem is responsible for absorbing the byproduct heat from combustion and transforming it into usable heat, thus improving the overall system efficiency. The first stage of the heat recovery subsystem is the engine cooling system. It consists of a simple volumetric pump, a splitter and a mixer. The coolant that flows through that circuit absorbs the engine heat in order to keep the engine temperature between certain bounds for efficiency purposes. The engine coolant interacts with a cold stream of usable water through a heat exchanger (Stage 1 interaction) in order for (i) the coolant temperature to decrease, (ii) the water temperature of the outer cooling system to increase, thus transforming the byproduct heat into usable heat. The Stage 2 interaction takes place in a second heat exchanger where the hot exhaust gases interact with the water of the outer cooling system in order to further increase the temperature of the latter.

The full length description of the high fidelity model as well as a series of simulation results is available in Diangelakis et al. (2014) [377]. The resulting dynamic model is implemented in the gPROMS environment. The complex DAE system features 379 equations (15 differential), and 6 degrees of freedom (4 dynamic).

System identification. According to the PAROC Framework, the high fidelity model is in some cases simple enough to be used for optimization and control purposes as is. The complexity of a DAE system, such as the one that describes this process, poses great challenges when the design of a model based controller is the objective, therefore model approximation is of the essence. In this section, we present the procedure of reducing the complexity of the DAE CHP model via the use of the System Identification Toolbox of MATLAB. The objective is the acquisition of discretized linear state-space model that adequately describes the systems behaviour with minor compromisation in terms of accuracy and response.

For this case study, a decentralized approach is used. The issue of decentralization is an open issue in terms of the application of model predictive control. Although it has been visited in the past [383, 384] we recognize that a straightforward approach is a matter that requires further research. Specifically for the CHP case study, the rationale behind the choice of a decentralized approach lies within the general concept of having a single equipment and the ability to produce two distinct products. More specifically, in the case of the CHP, the equipment (internal combustion engine, manifolds, heat recovery system and the like) is able to primarily produce electrical power to cover an electricity demand or primarily increase the temperature of water for heating purposes, thus covering a heating load demand. From this statement the following can be infered, (i) the same equipment can use two distinct "recipes" to alter its operation, thus (ii) producing products that cover different demands (i.e. different objectives). The decentralization approach in this case focuses on increasing the flexibility of the system to alternate its operation according to a criterion that affects the system only in terms of revenue but not of optimal operation. Since the two distinct products correspond to the operation of clearly defined subsystems of the equipment, the decentralization follows a similar path.[4] The two subsystems are treated separately which results into the generation of two state-space models. A series of input–output (I/O) data is introduced in the System Identification Toolbox, covering the entire range of values. Based on the (I/O) data, a state-space model is approximated.

---

[4] Although we mention two distinct products/objectives it is clear from the operation of any CHP system that one cannot be created without the other. Therefore, in terms of Model Predictive Control, the operation of one cannot be entirely decoupled from the other.

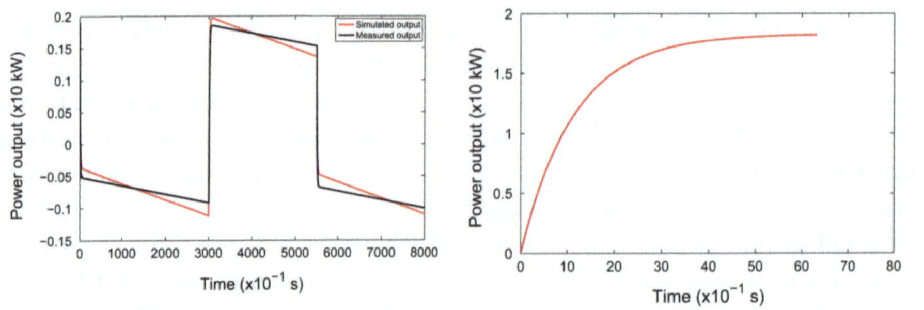

(a) Response of the power generation subsystem based on the high fidelity model and the approximate model.

(b) Step response of the approximate model for the power generation subsystem.

Fig. 3.6 Performance metrics for the approximate model of the power generation subsystem [227]

In the case of the power generation subsystem, a Single Input–Single Output (SISO) set of data is introduced. The input variable is the position of the butterfly in the throttle valve, with values that range between 0 and 1 and the output variable is the amount of electrical power generated by the CHP system. The identification procedure results into the generation of the discretized state-space model as follows.

$$x_{k+1} = 0.9913x_k + 0.0044u_k, \quad y_k = 3.5927x_k, \quad T_s = 0.1s \tag{3.8}$$

The state-space model does not account for any measured or unmeasured disturbances and, due to the identification process followed, the identified state cannot be correlated to a physical variable of the model.[5] The discretization time has been chosen to be 0.1 s with the controller design in mind. More specifically, an energy system is required to have fast dynamics and to be able to accurately account for rapid change. Consequently, in a discrete mpMPC formulation, and with the output and control horizon length in mind, a rather small discretization time should be considered. Figure 3.6a, b present the identified model mismatch (90.2%[6]) and its step response, respectively.

A similar procedure is followed for the identification of the heat recovery subsystem. The Multiple Input–Single Output (MISO) system is inserted in the System Identification Toolbox. In this case, the two inputs are the water mass flow rate in the inlet of the heat recovery subsystem and the power output of the power generation subsystem. The subsystem output is the water temperature at the outlet of the heat recovery subsystem. For consistency purposes, the discretization time is kept the same. The resulting 3 × 3 state-space model is presented in Eq. 3.9 and the identified system performance in Fig. 3.7a, b.

---

[5] In contrast to a variety of model reduction processes where the reduced model corresponds to physical variables (see Rivotti et al., 2012 [299]).

[6] As obtained by the System Identification Toolbox in MATLAB.

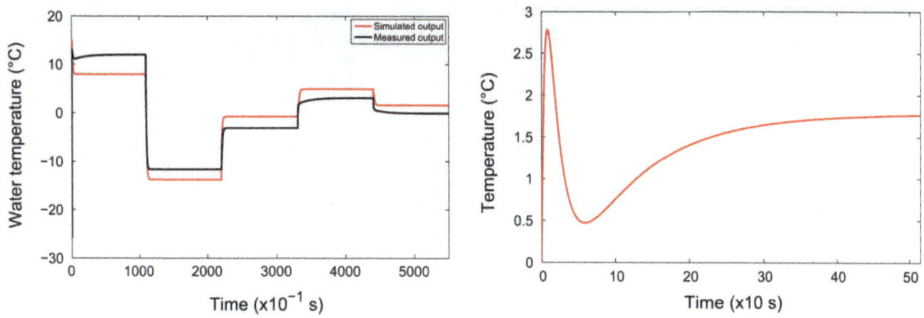

(a) Response of the heat recovery subsystem based on the high fidelity model and the approximate model.

(b) Step response of the approximate model for the heat recovery subsystem.

Fig. 3.7 Performance metrics for the approximate model of the heat recovery subsystem [227]

$$x_{k+1} = \begin{bmatrix} 0.9712 & -0.0207 & -0.0529 \\ 0.0012 & 0.8169 & -0.0524 \\ -0.0099 & -0.0302 & 0.9551 \end{bmatrix} x_k + \begin{bmatrix} -0.0245 & -0.0079 \\ -0.1009 & 0.0593 \\ -0.02457 & 0.0125 \end{bmatrix} u_k \tag{3.9}$$
$$y_k = \begin{bmatrix} -158.5 & -9.306 & -155 \end{bmatrix} x_k, \quad T_s = 0.1$$

A third state-space model is derived based on a continuous ordinary differential equation, that describes in a single state the water mass flow rate at the outlet of the heat recovery subsystem using the mass flow rate at the inlet of the system as its single input.

$$x_{k+1} = 0.99x_k + 0.01u_k, \quad y_k = x_k, \quad T_s = 0.1s \tag{3.10}$$

According to the PAROC Framework, the next step of the procedure is the design of the controller and its solution via multi-parametric programming.

Multi-parametric programming and mpMPC. In this section, the design of the controller and its solution via multi-parametric programming is presented. Every case though requires special attention and minor alterations in the PAROC Framework while at the same time keeping the general procedure intact. In this case study, the decentralized approach requires the development of a controller interaction scheme that enables the interconnection of the controllers in a manner that will (i) fit the systems characteristics and (ii) achieve the desired operational setpoints. In this case, the operational objective is the production of water of a predefined mass flow rate and temperature. The objective is chosen based on the fact that the usage of a domestic/residential CHP unit targets primarily heating rather than electricity production. Therefore, in this work, we choose to focus on satisfying heating demand while treating electricity as the byproduct of the CHP operation.

Three controllers are developed based on the identified systems. For brevity, the controllers based on the power generation and heat recovery subsystems are described, while

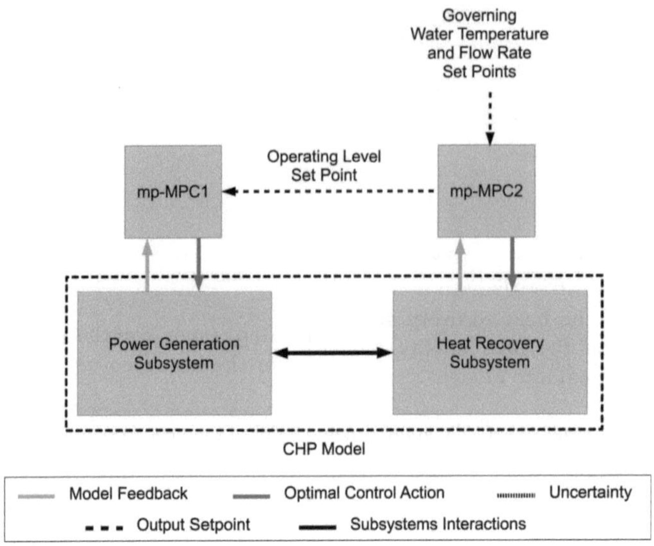

Fig. 3.8 Control scheme of the CHP system [227]

the controller of the mass flow rate is omitted. The heat recovery subsystem controller (mpMPC2) treats the water mass flow rate at the inlet of the heat recovery subsystem as a disturbance and the power production setpoint as a manipulating variable. The water temperature at the outlet of the system is the control variable. The manipulating variable of mpMPC2 becomes the operational setpoint of the power generation subsystem controller (mpMPC1). The manipulating variable of mpMPC1 is the throttle valve position. Figure 3.8 is a graphical representation of the decentralized control policy and the CHP system interactions.

The controllers of the systems are designed following standard multi-parametric MPC techniques [83] in MATLAB, using the POP toolbox[376]. The parameters of the quadratic optimization problems consist of the initial values of the states, the measured disturbances, the previous control action, the output setpoints and the output setpoint mismatch against the original high fidelity system. The mpMPC1 is solved for a control and output horizon of 10 time steps and features 4 bounded parameters while a significantly shorter horizon is chosen for mpMPC2 since the consideration of measured disturbances poses a great challenge in terms of computational expense, as the horizon increases. The total number of parameters for mpMPC2 is 8. In Fig. 3.9a, b 2-D projections of parametric critical regions for mp-MPC1 and mp-MPC2 are presented.[7]

[7] The values of the rest of the parameters are set to certain values within their feasible bounds in order to project a multi-dimensional solution space to 2-D.

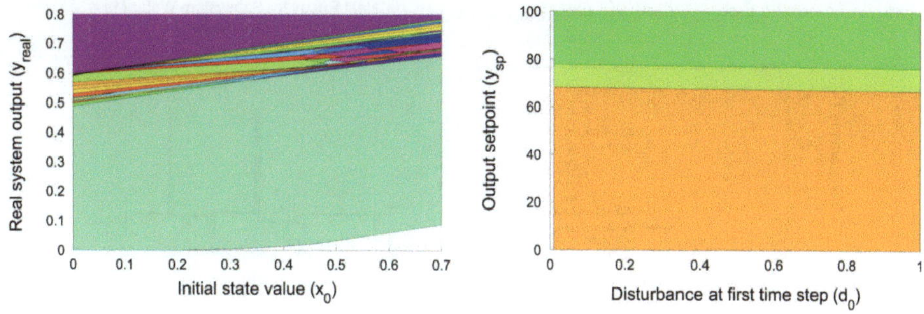

(a) Power generation subsystem mpMPC. (b) Heat recovery subsystem mpMPC.

Fig. 3.9 Partitions of the offline closed-loop control strategies projected on a 2-D parameter space [227]

Closed-loop validation. The last step of the PAROC Framework is the closed-loop validation, i.e. the testing of the controller against the original high fidelity model developed in gPROMS. The closed-loop validation is an important procedure since it verifies the accuracy and robustness of the controller. Depending on the outcome of the procedure, the controller design is evaluated.

In this case, the controller scheme is tested against the high fidelity model described formerly through gO:MATLAB, which enables the interconnectivity of gPROMS and MATLAB. The results of the closed-loop validation are presented in Fig. 3.10.

The closed loop validation results show that all the three mpMPC controllers manage to accurately follow the respective setpoints and operational objectives. More specifically mpMPC1 manages to accurately follow the power output setpoint provided by mpMPC2. The latter is able to effectively reject the disturbances in order to maintain the temperature of the water at the outlet of the heat recovery subsystem in a predefined setpoint. After the settling time the mismatch between the output setpoint and the real output is less than 2%. The flow controller manages to effortlessly maintain the water flow rate at the desired levels.

3.5 Concluding Remarks

In this work we developed the PAROC framework, a comprehensive approach towards the design, operational optimization and model-based control of process systems. Using the advances in the different fields of multi-parametric programming and model predictive control, the PAROC framework decomposes the problem into a series of steps. In the first step, a high-fidelity model of the process is developed and sensitivity analysis, parameter estimation and dynamic optimization can be applied. In the second step, the initial high-fidelity model is reduced using system identification as well as recently developed model-reduction techniques. This reduced model is used in the next step to obtain the explicit

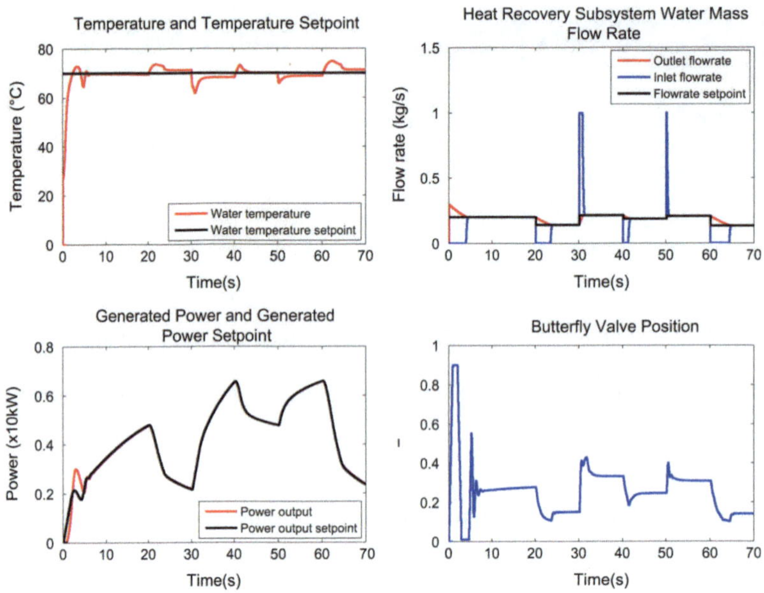

Fig. 3.10 Closed loop validation results [227]

control law of the system, which is calculated once and offline. This control law is then validated in-silico against the original high-fidelity model, thus closing the loop.

The key advantages of PAROC are thereby (i) its effortless applicability on fundamentally different classes of problems, (ii) its decomposed nature which allows for different advanced applications in the different steps, (iii) the inherently offline solution of the problem thus limiting the online computational burden and (iv) the validation of the exact solution derived from the simplified model against the high-fidelity model through the interconnectivity of the different software packages.

Furthermore, we presented a case study with a residential scale heat and power cogeneration unit which elucidated different aspects of the framework. We designed a decentralized control scheme to satisfy the heat demand. This scheme decomposes the original system into two major parts, enabling a more effective approach. In the following chapters, we will discuss how to use this framework to integrate process design and operational decisions based on a single high fidelity model.

4 Integrating Process Design Optimization and Advanced Model Based Control Strategies

In this chapter, we extend the PAROC framework introduced in Chap. 3 to optimize the process design, simultaneously with the advanced control strategy. We develop design dependent multi-parametric model predictive controllers that are able to provide the optimal control actions as functions of the system state and the design of the process at hand. The process and the design dependent explicit controllers undergo a Mixed Integer Dynamic Optimization (MIDO) step for the determination of the optimal design. The result of the MIDO is the optimal design of the process under optimal operation. We first demonstrate the framework on the illustrative CSTR example introduced in Chap. 3, and then explore it further on more complex applications, namely a binary distillation column, an intensified reactive distillation column, and a residential cogeneration unit [85, 385].[1]

4.1 Overview

Integrating dynamic control actions into the time-invariant design decisions has been explored in the PSE community for more than three decades in order to address the issue of operability during the early stages of process design. Process design optimization under operational uncertainty and feasibility, flexibility, stability, controllability and resilience metrics during process design have been extensively discussed via a series of computational methods (e.g. [44, 386]). This formed a prelude to the simultaneous consideration of design

[1] Reprinted from AIChE Journal, vol. 63, no. 11, N. A. Diangelakis, B. Burnak, J. Katz, E. N. Pistikopoulos, Process design and control optimization: A simultaneous approach by multi-parametric programming, pp. 4827–4846, 2017 and Computers and Chemical Engineering, Y. Tian, I. Pappas, B. Burnak, J. Katz, and E. N. Pistikopoulos, A systematic framework for the synthesis of operable process intensification systems—reactive separation systems, vol. 134, 106675, 2020, with permission from Elsevier [85, 385].

B. Burnak et al., *Integrated Process Design and Operational Optimization via Multi-parametric Programming*, Synthesis Lectures on Engineering, Science, and Technology, https://doi.org/10.1007/978-3-032-10181-5_4

and control via (i) the formulation and solution of large scale optimization problems (including numerous decomposition approaches, e.g. [387]), (ii) flowsheet and graphical problem representations (e.g. [388]) and (iii) control structure selection as part of the design optimization (e.g. [389]) (see Table 4.1 for a list of publications per contribution). The control schemes employed focused mainly on PI and PID formulations while a significantly smaller portion of contributions employed Model Predictive Control (MPC). The contributing factor to that decision was primarily the solution of the optimization problem corresponding to the control problem within a design optimization formulation (e.g. [35, 71]). Nevertheless, the consideration of a constrained optimization control method could contribute to overcome the shortcomings associated with PI and PID control (such as possible operational constraints violation). In the area of simultaneous design optimization with MPC notable approaches include (i) the back-off control approach (e.g. [390]), (ii) robust design formulations (e.g. [391]) and (iii) multi-parametric MPC approaches (e.g. [84]) (see Table 4.1 for a list of publications per contribution). Regarding (iii), the availability of the optimal solution online via offline optimization enabled the incorporation of explicit control actions within a (Mixed Integer) Dynamic Optimization ((MI)DO) formulation thus (i) avoiding the burden of solving multiple optimization problems online, (ii) transforming the control problem into a simple linear look-up function[2] and (iii) including every aspect of the MPC without any simplifications on the problem structure.

Although multi-parametric Model Predictive Control (mpMPC) has been employed in the past in the context of simultaneous design and control optimization [84, 420–422], its application relied on an iterative procedure, because the control problem formulation needed to be adjusted for different design alternatives based on feasibility criteria. Here, we present a methodology, via the PAROC framework and software platform (see Chap. 3), where the control problem formulation is design dependent, therefore, the explicit control actions are a function of the design variables. As a result, a single design dependent mpMPC formulation is able to control the process for bounded values of the design variables without the need of reformulation. The approach is showcased via four case studies on (i) a continuously stirred tank reactor, (ii) a binary distillation column, (v) and intensified reactive distillation column, and (iv) a domestic cogeneration of heat and power unit.

4.2 Simultaneous Design and Control Optimization via PAROC

In Chap. 2 we defined a general mathematical model for a process based on first principles and correlations as an MIDO problem, given by Eq. 2.1. In this section, we will tailor the PAROC framework to develop a single design-dependent mpMPC, which will be incorporated in Eq. 2.1 in its explicit form.

[2] The explicit solution of a Model Predictive Control problem with a linear ($\infty - norm$) or quadratic ($2 - norm$) objective function, polytopic constraints and linear state-space discrete time model dynamics is piecewise linear in the optimal actions [83, 392].

Table 4.1 An indicative list of integrated process design and control approaches in the literature [85]

Author (year)	Contribution
Perkins and co-workers (1991) [59], Bogle and co- (1989, 2000) [393, 394], Pistikopoulos and co- (1994, 1997, 2001) [36, 395, 396], Floudas and co- (1994, 2000, 2001) [397–399], Romagnoli and co- (1997) [400], Ricardez-Sandoval and co- (2007, 2013) [401, 402], Douglas and co- (1988) [403–405], Skogestad and co- (1987, 2014) [406, 407], Sorensen and co- (2014) [408], Stephanopoulos and co- (1988) [409], Ierapetritou and co- (2002) [410], Gani and co- (1995) [411]	Feasibility, flexibility, stability, controllability and resilience considerations in steady-steady state [w/wo (MI)NLP design optimization]
Romagnoli and co- (1996) [412], Francisco and co- (2014) [413], Kravaris and co- (1993) [414, 415]	Feasibility, flexibility, controllability and resilience considerations in steady-steady state (MI)DO design optimization
Pistikopoulos and co- (2000, 2002, 2003) [71, 228, 416–418], Swartz and co- (2014) [419], Ricardez-Sandoval (2012) [386]	Simultaneous/Decomposition (MI)DO process and P-PI-PID control design
Pistikopoulos and co- (2003, 2004) [84, 420–422], Engell and co- (2004) [423], Linninger and co- (2007) [80]	Simultaneous/Decomposition (MI)DO process and MPC design
Biegler and co- (2007, 2008) [387, 424], Seider and co- (1992) [82], Ricardez-Sandoval and co- (2008, 2016, 2017) [72, 77, 425], Pistikopoulos and co- (1996) [35], Perkins and co- (2002, 2004, 2016) [75, 75, 426, 427], Flores-Tlacuahuac and co- (2009) [428], Barton and co- (2010, 2011, 2015) [429–432], Mitsos and co- (2012) [433], Linninger and co- (2006) [434]	Simultaneous/Decomposition/Back-off via (MI)NLP
Gani and co- (1995, 2003, 2005, 2010) [388, 411, 435–437], Daoutidis and co- (2011) [438], Lee and co- (1972) [439], Chien and co- (2010) [440], Mitsos and co- (2014) [441], Luyben (2004, 2008, 2009, 2010, 2011, 2012, 2014) [442–455]	Flowsheet/Graphical design and P-PI-PID control
Floudas and co- (1994) [397], Pistikopoulos and co- (1997) [456], You and co- (2012) [457]	Multi-objective approaches
Floudas and co- (2001) [399], Ierapetritou and co- (2002) [410], Barton and co- (2010, 2015) [429, 432], Ricardez-Sandoval and co- (2013, 2015) [4, 87], Pistikopoulos and co- (1995, 1999, 2000, 2003) [44, 458–460], McRae and co- (2007) [461], Bogle and co- (2006) [462]	Design under uncertainty
Skogestad and co- (1987, 1989, 2014) [389, 407, 463], Perkins and co- (2002) [426], Young and co- (2005) [464], Stephanopoulos and co- (1980, 1988) [465, 466], Bogle and co- (2010, 2016) [467, 468]	Control structure selection and design
Georgiadis and co- (2004) [469], Francisco and co- (2014) [470], Ricardez-Sandoval and co- (2009, 2011) [471, 472], Gani and co- (2012) [473], Mitsos and co- (2014) [474]	Review articles on design and control
Ricardez-Sandoval and co- (2022) [94]	Integration of design and nmpc-based control under uncertainty framework

Incorporating Process Design in the PAROC Framework

Here, we extend the PAROC framework to develop a single design dependent mpMPC strategy that can be used for the entire range of design parameters, which will be incorporated

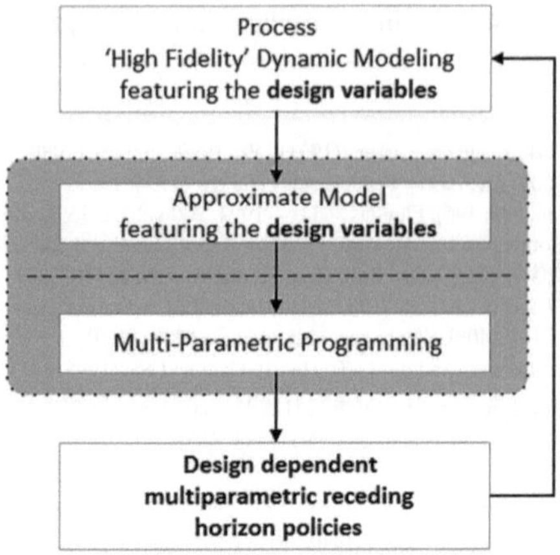

Fig. 4.1 Simultaneous process design and control using the PAROC framework [85]. Actions within the gray area happen once and offline

in an MIDO for simultaneous design and control. The two major differences from the generic framework introduced in Chap. 3 are (i) a design dependent approximate model, and (ii) the final MIDO step to determine the process design. Therefore in this chapter, we will focus on these steps. The remaining steps are exactly the same, and summarized in Fig. 4.1.

Model approximation. For simultaneous design and control, the key step is to develop a single explicit control strategy that can be used for the entire range of design options. Therefore, a high fidelity model described by a differential algebraic equation system (DAE) (Eq. 4.1) can be approximated by a design dependent state space model given by 4.2.

High fidelity model

$$\begin{aligned} \dot{x}(t) &= f(x(t), u_c(t), Y(t), d(t), De) \\ y &= g(x(t), u_c(t), Y(t), d(t), De) \end{aligned} \tag{4.1}$$

Approximate model

$$\begin{aligned} x_{T+1} &= Ax_T + Bu_{c,T} + C\begin{bmatrix} d_T \\ De \end{bmatrix} \\ y_T &= Dx_T + Eu_{c,T} + e \end{aligned} \tag{4.2}$$

In Eq. 4.2, the design of the process is treated as a measured additive uncertainty within the approximate model via the term De. Multiplicative uncertainty can be incorporated without any changes in the framework. This can be achieved via the consideration of robust multi-parametric Model Predictive Control as described in [360, 362, 475]. In such case, the use of state-estimators is also necessary. Note that De are not time dependent since the design of the process cannot change during operation.

Multi-parametric programming and closed loop validation. The design dependent approximate model is used as the governing model in the mpMPC formulation given by Eq. 3.1. The solution of the mpMPC yields the explicit optimal control law given by Eq. 4.3.

$$\begin{aligned} u_{c,T} &= K_i \theta_T + r_i, \ \forall \theta_T \in CR_i \\ \theta_T &= [x_T; u_{c,T-1}; d_T; De; y_T; y_T^{SP}] \end{aligned} \tag{4.3}$$

Equation 4.3 is embedded in the dynamic high fidelity model Eq. 4.1 and is validated by subjecting the integrated model to a series of disturbance profiles at different design configurations.

Dynamic optimization. Through the creation of Dynamic Link Libraries the design dependent control scheme is introduced into gPROMS®. Problem 2.1 is therefore reformulated as in Eq. 4.4.

$$\begin{aligned} \min_{u_{c,T}, Y, De} \quad & J = \int_0^{\tau} P(x, y, u_c, Y, d, De) dt \\ \text{s.t.} \ & \dot{x} = f(x, u_c, Y, d, De) \\ & y_{\min} \le y = g(x, u_c, Y, d, De) \le y_{\max} \\ & u_{c,T} = K_i(\theta_T) + r_i, \ \forall \theta_T \in CR_i \\ & \theta_T = [x_T; u_{c,T-1}; d_T; De; y_T; y_T^{SP}] \\ & Y \in \{0, 1\} \\ & [x_{\min}^T, d_{\min}^T]^T \le [x^T, d^T]^T \le [x_{\max}^T, d_{\max}^T]^T \\ & De_{\min} \le De \le De_{\max} \end{aligned} \tag{4.4}$$

Problem 4.4 is a mixed integer dynamic optimization program which is handled via a control vector parameterization (CVP) algorithm in gPROMS®. The term T denotes the control time step interval and τ the horizon of the MIDO problem. The time-varying optimization variables are piecewise-constant functions of time over a number of intervals which are specified by the user according to the needs of each problem. Note that the duration of the intervals in each step are in this case determined by the user to be equal in duration with the control time step for synchronization purposes.[3] The single vector shooting dynamic optimization can be decomposed as follows[4]:

- The values of the optimization variables are determined for each interval.
- The dynamic model is simulated over the entire time horizon.
- The value of (i) the objective function and its partial derivatives with respect to the optimization variables and (ii) the constraints are determined.
- Convergence is checked and if needed the steps are repeated.

[3] In the general case this is an extra degree of freedom for the dynamic optimizer.

[4] For more information the user is referred to the Optimization Guide of PSE's gPROMS®.

According to the Optimization Guide of PSE's gPROMS® the CVP algorithm for (MI)DO relies on the values of the partial derivatives of the problem's objective function with respect to the optimization variables to determine its next iteration step. In the case of the design optimization variables, the partial derivatives are needed with respect to the rest of the 'high-fidelity' model variables (x and y) as well as the control variables ($u_{c,T}$). Although the derivatives of the former are calculated "on the fly" via the optimization algorithm, the calculation of the derivatives of the latter can become an issue when an external piece of software introduces the $u_{c,T}$ values to the problem. This burden is alleviated with the use of multi-parametric Model Predictive control as the partial derivatives of the control actions with respect to the design variables (optimization variables in the (MI)DO context) are available *a priori* via K_i in Problem 4.4 as exact expressions, not numerical approximations, thus utilizing in full the concept of "map of solutions" introduced in [360]. Also note that the dynamic model simulation in gPROMS® happens based on non-uniform discretization steps posing a technical, time synchronization challenge to the overall simultaneous approach. In order to overcome this, an error function is defined for every control action which absorbs the evaluation of the controller at the non-uniform time steps of the simulation and only utilize the evaluation at the control time intervals. This happens via imposing interior point constraints of the form $err(t) = u_{c,T} - u(t) = 0$. The latter is zero only at every interior point of the MIDO. The variable $u(t)$ therefore remains piecewise constant, between interior points, and equal to the value of $u_{c,T}$ at the interior point, although from a software point of view the variable $u_{c,T}$ is free to be evaluated multiple times throughout a single MIDO time step. The evaluation of $u_{c,T}$ within the MIDO time step is therefore rejected. Also note that within the MIDO problem, the optimal control action $u_{c,T}$ is regarded as an optimization variable although it is calculated via the mpMPC formulation of Step 3. The two reasons that this is happening is (i) the fact that the action is implicitly optimized via the optimal choice of the design variables De and the dependence of $u_{c,T}$ to those variables and (ii) the fact that within the MIDO problem formulation this is necessary in order to achieve the synchronization of the fixed time-step control with the MIDO problem.

Based on the nature of the design dynamic optimization model and especially the existence of binary variables, suitable optimization solvers are employed for the task of optimizing the model at every interval. More specifically, for the general case of MIDO an Outer approximation based method is employed where the problem is initially reformulated to a completely relaxed NLP. A linearized version of the model excluding any binary variable combinations from previous iterations is subsequently solved (master problem). Based on the integer solution of the master problem, the primal problem is formulated and solved. Both the primal and the fully relaxed problem are solved via a sequential quadratic programming algorithm (SQP). Note that in this work we use a commercially available software tool for the solution of the design (MI)DO problem. Based on the aforementioned characteristics of the solver the solution is guaranteed to be locally optimal, even for the DO case. For a list of available MIDO algorithms see Table 4.2. The overall simultaneous design and control optimization is schematically presented in Fig. 4.2. The dynamic optimization algorithm utilizes

Table 4.2 An indicative list of (MI)DO algorithms in literature [85]

Author (year)	Contribution
Pistikopoulos and co-workers (2003) [228]	Algo. 1: GBD based approach. The master problem is constructed without the solution of an intermediate adjoint problem at the expense of including additional equations and search variables
	Algo. 2: Master problems are computationally more expensive due to addition of linearizations about the optimal solution of the primal problem. Additional constraints render the master problem tighter, hence fewer iterations suffice for a local solution
Barton and co-workers (2006, 2009) [476, 477]	Algo. 1: Outer approximation based approach for global optimization. Primal problem is not solved to global optimality at every iteration, which alleviates the computational burden
	Algo. 2: Formulates a bilevel dynamic optimization problem, which can be naturally extended to a mixed-integer dynamic optimization formulation. A branch and bound algorithm is utilized to solve the problem to global optimality
Biegler and co-workers (2007) [387]	Full discretization by finite elements, and solving a large scale nonconvex MINLP. Generalized disjunctive programming are used to solve the MINLP
Marquardt and co-workers (2008) [478]	MIDO problem is reformulated as a mixed-logic dynamic optimization problem, and solved by control vector parameterization and direct multiple shooting
You and co-workers (2013, 2014, 2015) [92, 93, 479–482]	Algo. 1: The inner level dynamic optimization problem is replaced with a set of surrogate models, which are updated adaptively with every iteration
	Algo. 2: MIDO is reformulated as a large scale MINLP problem. MILP master problem is subjected to a bilevel decomposition algorithm based on the inherently different time scales of the original problem
	Algo. 3: GBD based approach. Decomposed primal problem is a set of separable dynamic optimization problems, and the master problem is a mixed-integer nonlinear fractional problem, which is solved to global optimality by a fractional programming algorithm

information from the process and the optimal control actions derived multi-parametrically to determine the optimal design. The values for the optimal design are used to calculate the numerical values of the control actions and progress the simulation step.

The following sections present the application of the simultaneous design and control framework via a tank, a CSTR, a binary distillation column, a reactive distillation column, and a domestic cogeneration unit examples.

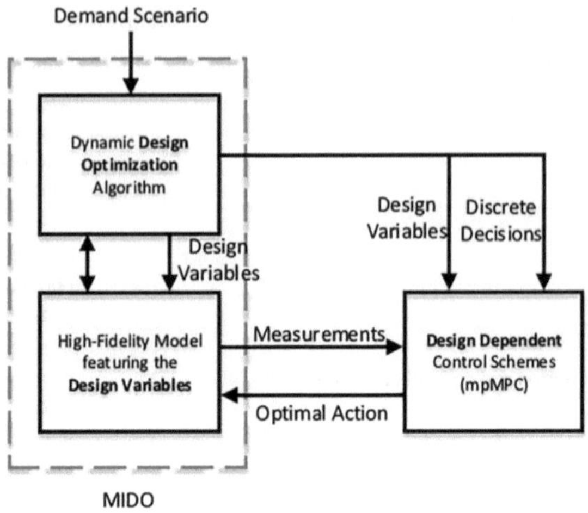

Fig. 4.2 Interactions between the decision layers. The area within the dashed line represents the MIDO problem [85]

4.2.1 Illustrative Example

In this example, we develop on the CSTR example introduced in Sect. 3.3 to simultaneously account for the design decisions. The volume of the CSTR is an essential decision that impacts the closed loop dynamics, and hence needs to be accounted for simultaneously with the mpMPC development. Therefore, we use the high fidelity model from the previous section as the basis. However, since the design decisions interact with the other input, i.e. reactant flow rate Q, we develop a new approximate state space model using the MATLAB System Identification Toolbox. The developed approximate model is given in Eq. 4.5.

$$
\begin{aligned}
A &= \begin{bmatrix} 0.86 & 0.00 & -0.06 & 0.13 & -0.05 & 0.14 & -0.33 & 0.60 \\ 0.00 & 1.00 & 0.01 & -0.02 & -0.02 & -0.12 & 0.53 & -0.84 \\ -0.01 & 0.00 & 0.95 & 0.33 & -0.10 & -0.06 & 1.09 & -1.62 \\ -0.01 & 0.01 & -0.06 & 0.60 & 0.5600 & -0.46 & -0.17 & -0.47 \\ 0.00 & 0.00 & 0.00 & -0.02 & 0.85 & 0.64 & 0.26 & 0.09 \\ 0.00 & 0.00 & 0.00 & 0.01 & 0.05 & 0.73 & -0.92 & -0.19 \\ 0.00 & 0.00 & 0.00 & 0.00 & 0.00 & -0.01 & 0.79 & -1.10 \\ 0.00 & 0.00 & 0.00 & 0.00 & 0.00 & 0.00 & 0.00 & 0.2900 \end{bmatrix} \\
B \times 10^3 &= \begin{bmatrix} -2.8 \\ 0.3 \\ -3.6 \\ 6.4 \\ -2.3 \\ 1.7 \\ 0.6 \\ 0.2 \end{bmatrix}, \quad C \times 10^3 = \begin{bmatrix} 0.2 \\ -0.0 \\ 0.3 \\ -0.3 \\ -0.0 \\ 0.1 \\ 0.1 \\ 0.0 \end{bmatrix} \\
D &= \begin{bmatrix} 0.80 & 4.06 & -0.08 & 0.02 & -0.06 & -0.00 & -0.00 & 0.00 \end{bmatrix}
\end{aligned}
\tag{4.5}
$$

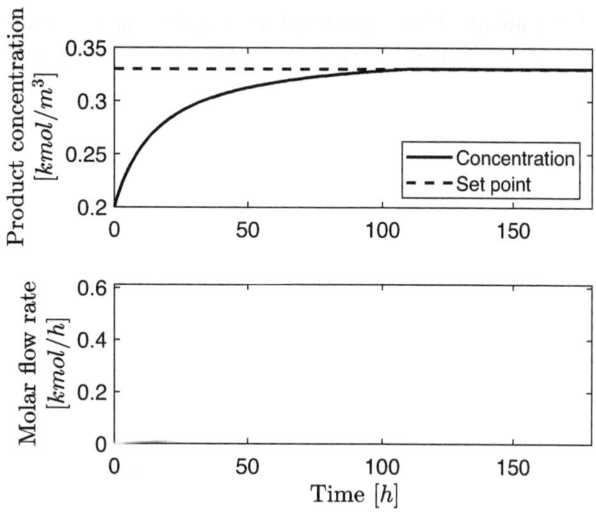

Fig. 4.3 A snapshot of the closed loop profile of the CSTR

Note that Eq. 4.5 has two inputs, namely the reactant flow rate and the reactor volume. However, only the former is considered as a manipulated variable and the latter is a bounded parameter in the mpMPC formulation, Eq. 3.1. The controller is validated rigorously against the high fidelity model at different design configurations, and embedded in the dynamic optimization formulation in the gPROMS environment. The capital cost of the reactor is given by Eq. 4.6.

$$C_e = a + bV^n \tag{4.6}$$

where C_e is the annualized reactor cost, and a, b, and n are cost parameters given in Appendix D.3 with the cost escalation indexes for year 2018. Note that the set point for the controller is an optimization variable in the MIDO and a parameter only in the mpMPC formulation. The MIDO is solved for 4 h and the optimal design configuration is found at $V = 0.78\ \text{m}^3$. A snapshot of the closed loop profile is presented in Fig. 4.3.

Binary Distillation Column

The distillation column model describes the binary separation of benzene and toluene. The column is allowed a maximum number of trays to be 30 with no restriction on feed tray location. The purity in the top has a desired set point of 0.98 and the purity in the bottom has a desired setpoint of 0.02. The feed composition is assumed sinusoidal and is optimized similarly to the previous two examples.

High fidelity dynamic modelling. The distillation column utilizes mass and energy balances and thermodynamic relations to build the full model. The following assumptions have been made:

- Fast energy dynamics
- Relative Volatility
- Constant molar hold up in condenser
- Immediate pressure response

Mass balances for each tray, reboiler, and condenser are used while assuming constant molar hold up in the total condenser. Energy balances are used in the reboiler and condenser while assuming an average temperature throughout the column. Relative volatility is used to determine vapor and liquid correlations in each tray and in the reboiler. The model assumes the reflux flow rate and the boil up rate to be the controllable variables in the system, and the molar hold ups to be the states of the system. Column diameter, reflux tray position, and feed tray position are the design variables, while the presence and position of the reboiler and condenser are fixed. Density of the liquid hold up on the trays is assumed to follow from a linear combination of the component densities.

The model of the binary distillation column is adapted from [84]. The characteristic equations of the model are presented in Table 4.3 and in Appendix B.1, Table B.1 for the nomenclature.

Antoine equations were used to determine the vapor pressures at the top and bottom of the distillation column and the log-mean temperature approach was used for the heat exchange at the condenser and the reboiler. Since the 'high-fidelity' model used here is a simplified model, the limiting constraints of the columns operation are expressed via thermodynamic limits which are manifested via the Antoine equations and subsequently the relative volatility relations. The distillation column is a multiple input multiple output (MIMO) system where the reflux flow rate and boil up flow rate are the degrees of freedom to the system, and the purity in the top and bottom is the output. The composition at the feed is treated as a disturbance to the system operation.

Model approximation. The high fidelity model of the distillation column consists of 50+ states and nonlinear equations. Random sets of I/O for different designs from the 'high fidelity' model are introduced into the System Identification Toolbox in MATLAB® to acquire a linear state-space model of the form of Eq. 4.2. The identified state-space model is shown in Eq. 4.7.

Table 4.3 High fidelity model of the binary distillation column example [85]. Component i is benzene, tray number is $k \in \{1...Ntrays\}$ unless stated otherwise

Description	Equation
Component mass balance	$\frac{dM_{i,k}}{dt} = L_{k+1}x_{i,k+1} + V_{k-1}y_{i,k-1} + F_k z_{i,f} + R_k x_{i,D} - L_k x_{i,k} - V_k y_{i,k},\ \forall k \in \{2...Ntrays - 1\}$
Total mass balance	$\frac{dM_k}{dt} = L_{k+1} + V_{k-1} + F_k + R_k - L_k - V_k,\ \forall k \in \{2...Ntrays - 1\}$
Vapor molar flow rate	$V_k = V_{k-1} = V_B,\ \forall k \in \{2...Ntrays - 1\}$
Hold-up	$Vol_k = \frac{M_k}{\rho_{Lmix,k}}$
Liquid level	$Level_k = \frac{L_k^{2/3}}{1.84\rho_{Lmix,k}L_{weir}} + H_{weir}$
Tray area	$A_{tray} = \frac{0.8\pi D_c^2}{4}$
Weir length	$L_{weir} = 0.77D_c$
Reboiler vapor liquid equilibrium	$1 = \frac{P^0_{benz,B}x_{i,B} + P^0_{tol,B}(1-x_{i,B})}{P}$
Condenser vapor liquid equilibrium	$1 = P(\frac{x_{i,D}}{P^0_{benz,D}} + \frac{1-x_{i,D}}{P^0_{tol,D}})$
Relative volatility	$\alpha = \sqrt{\frac{P^0_{benz,D}P^0_{benz,B}}{P^0_{tol,D}}P^0_{tol,B}}$
	$y_{i,k} = \frac{\alpha x_{i,k}}{1+x_{i,k}(\alpha-1)}$
Reboiler and reflux drum molar balance	$\frac{dM_{i,B}}{dt} - L_1 x_{i,1} - Bx_{i,B} - V_B y_{i,B}$ $M_B = \frac{M_{i,B}}{x_{i,B}}$
	$\frac{dM_{i,D}}{dt} = V_{Ntrays}(y_{i,Ntrays} - x_{i,D})$
	$M_D = \frac{M_{i,D}}{x_{i,D}}$
Reboiler and reflux drum energy balance	$0 = L_1 - B - V_B$
	$0 = V_D - \Sigma R_k - D$

$$x_{T+1} = \begin{bmatrix} 0.9533 & -0.05507 \\ 0.0264 & 0.5494 \end{bmatrix} x_T + \begin{bmatrix} -0.01609 & -0.01346 \\ -0.1129 & 0.08987 \end{bmatrix} u_{c,T} + \tag{4.7}$$

$$\begin{bmatrix} -0.1257 & -9.70310^{-5} & -4.16310^{-4} \\ -1.005 & 7.18410^{-4} & -5.87410^{-5} \end{bmatrix} \begin{bmatrix} d_T \\ De \end{bmatrix}$$

$$y_T = \begin{bmatrix} -0.2357 & -0.354 \\ 0.1098 & -0.4719 \end{bmatrix} x_T \quad T_s = 1\,s$$

where x_T are the identified states, $u_{c,T}$ are the reflux flow rate and the boil up rate, d_T is the composition of the feed and De is the feed and reflux tray location. Note that the column diameter is correlated to the minimum vapor flow rate and therefore is the design decision of the system (See Eq. 4.8). Also note that the location of the trays are integer variables the handling of which in terms of multi-parametric programming will be discussed in the next section.

Design of the multi-parametric model predictive controller Similarly to the tank example in Sect. 4.2.1 the problem formulation is based on Eq. 3.1 and the tuning of the controller is presented in Table 4.4. Note that since the boil up flow rate is limited by the column diameter as presented in Eq. 4.8, the mpMPC is modified to account for the square of the column diameter as a design parameter. Note that since the column diameter is always greater than zero and it does not appear anywhere else within the mpMPC formulation we can define a new parameter $p = D_c^2$ which renders Eq. 4.8 a linear inequality constraint.

$$0.4514 V_B \leq D_c^2 \tag{4.8}$$

The integer parameters corresponding to the tray locations are reformulated into binary parameters and solved based on the algorithm presented in [483]. An alternative formulation could be the treatment of integer parameters as continuous parameters (similar to handling binary variables in [325]) since the integer value realization of the parameter is a subset of their continuous values and the realization is not an mpMPC decision.

The objective of the design dependent controller is to maintain the purity set points for the top and bottom product at 98% vol and 2% vol regardless of the disturbance at the inlet of the system. Small deviations are allowed but penalized in the design optimization formulation.

Closed-loop validation. The validation step can be seen in Fig. 4.4 where the controller is tested against the original high fidelity model. Note that the closed-loop validation needs to happen and be satisfactory for a range of different designs. Here we present the closed loop validation for a distillation column with the following characteristics:

- Condenser Area: 100 m^2
- Reboiler Area: 282.427 m^2
- Diameter of Column: 1.9 m

Table 4.4 Tuning parameters of the mpMPC for the distillation column example [85]

MPC design parameters	Value
N	3
M	1
$QR_k, \forall k \in \{1, \ldots, N\}$	$\begin{bmatrix} 10^7 & 0 \\ 0 & 10^7 \end{bmatrix}$
$R_k, \forall k \in \{1, \ldots, M\}$	$\begin{bmatrix} 10^{-2} & 0 \\ 0 & 10^{-2} \end{bmatrix}$
$x_{\min}$	$\begin{bmatrix} -10^3 & -10^3 \end{bmatrix}^T$
$x_{\max}$	$\begin{bmatrix} 10^3 & 10^3 \end{bmatrix}^T$
$u_{\min}$	$\begin{bmatrix} 2 & 3 \end{bmatrix}^T$
$u_{\max}$	$\begin{bmatrix} 4.7 & 7 \end{bmatrix}^T$
$y_{\min}$	$\begin{bmatrix} 0 & 0 \end{bmatrix}^T$
$y_{\max}$	$\begin{bmatrix} 1 & 1 \end{bmatrix}^T$
$d_{\min}$	$\begin{bmatrix} 0.45 & 1 & 1 \end{bmatrix}^T$
$d_{\max}$	$\begin{bmatrix} 0.5 & 30 & 30 \end{bmatrix}^T$

- Reflux Tray position: 18
- Feed Tray position: 9

Dynamic optimization. The dynamic optimization in Problem 4.4 is then formulated and solved allowing for the optimizer to select the optimal value for the area of the condenser, area of the reboiler, reflux tray location, feed tray location, and diameter of the column. To account for the reflux and feed tray location changing additional equations were added or modified as seen in Table 4.5. Allowing the dynamic optimization to run over a time

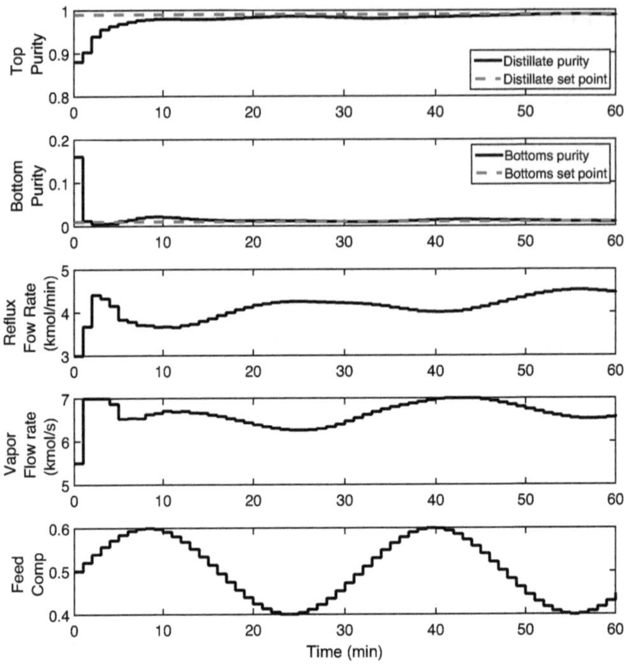

Fig. 4.4 Closed loop validation of the design dependent mpMPC against the high fidelity model [85]

span of 1 h, the results obtained are presented in Table 4.6. It can be seen that by utilizing the simultaneous design and control approach presented here, a distillation column with a smaller annualized total cost is designed.

Reactive Distillation Column

The objective of this example is to demonstrate the applicability of the framework on intensified processes. Herein, a reactive distillation column and its required explicit model predictive control scheme are designed simultaneously at minimum total annualized cost to maintain feasible operation in the presence of given disturbances over a finite time horizon of interest (i.e. 24 h).

The solution of this problem will give:

- the optimal RD column design and operating conditions—i.e. the number of column trays and feed tray location (discrete decisions), column diameter, reactive tray location with catalyst load (continuous decisions)
- the optimal design-aware explicit MPC controller design—i.e. the tuning parameters for mp-MPC.

Table 4.5 Additional/Modified equations for dynamic optimization [85]

Description	Equation
Feed tray location	$F_k = F\delta_k^f,\ \sum_{k=1}^{Ntrays} \delta_k^f = 1$
Reflux tray location	$R_k = R\delta_k^r,\ \sum_{k=1}^{Ntrays} \delta_k^r = 1$
Feed tray location only below reflux	$\delta_k^f - \sum_{k'=k}^{Ntrays} \delta_{k'}^r \leq 0$
Component mass balance	$(\sum_{k'=k}^{Ntrays} \delta_{k'}^r)\frac{dM_{i,k}}{dt} = L_{k+1}x_{i,k+1} + V_{k-1}y_{i,k-1} + F_k z_{i,f} + R_k x_{i,d} - L_k x_{i,k} - V_k y_{i,k},\ \forall k \in \{2\ldots, Ntrays-1\}$
Total mass balance	$(\sum_{k'=k}^{Ntrays} \delta_{k'}^r)\frac{dM_k}{dt} = L_{k+1} + V_{k-1} + F_k + R_k - L_k - V_k,\ \forall k \in \{2\ldots, Ntrays-1\}$
Reboiler cost	$C_{reb} = 0.6101.3\frac{M\&S}{280}(\frac{10^4 A_R}{1442.54^2})^{0.65}3.221.35$
Total cost	$TotalCost = OpCost + \frac{1}{3}(C_{column} + C_{reb} + C_{cond})$

Component i is benzene, tray number is $k \in \{1\ldots Ntrays\}$ unless stated otherwise

Table 4.6 Comparison of the results against Sakizlis et al. [84, 85]

	Proposed approach	Sakizlis et al. [84]
Condenser area $[m^2]$	120	132
Reboiler area $[m^2]$	266	276
Diameter of column $[m]$	1.62	1.65
Reflux tray	25	25
Feed tray	12	12
Total cost [k$]	590	620

Specifically, we investigate the production of methyl tert-butyl ether (MTBE) using reactive distillation. This process has been widely reported in the open literature with computational and experimental studies [484–488], which enables the validation of modeling predictions on this system as well as the benchmark of proposed approach for design and operational optimization.

In this reactive distillation system, MTBE is produced by reacting isobutylene (IB4) with methanol (MeOH) in the liquid phase catalysed by Amberlyst 15, an ion-exchange resin. N-butene (NB4) is also present in the reactive mixture as an inert compound. The reaction scheme is shown in Eq. 4.9, the intrinsic reaction rate of which can be accurately described with Eq. 4.10 adapted from Rehfinger and Hoffmann [489].

$$\text{MeOH} + \text{IB4} \rightleftharpoons \text{MTBE}, \quad \Delta_r H^o_{298\text{K}} = -37.7\text{kJ/mol} \tag{4.9}$$

$$r = k\left[\frac{a_{IB4}}{a_{MeOH}} - \frac{1}{K_a}\frac{a_{MTBE}}{a^2_{MeOH}}\right] \text{ kmol/(h} \cdot \text{kg cat)} \tag{4.10}$$

where r gives the molar reaction rate per unit mass of dry catalyst resin, a denotes the activity of each component. The rate constant k is given by Rehfinger and Hoffmann [489]:

$$k = 8.5132 \times 10^{13}\exp\left[\frac{-11,113.78}{T}\right] \text{ kmol/(h} \cdot \text{kg cat)} \tag{4.11}$$

where T denotes the process temperature in K. The expression of reaction equilibrium constant K_a is provided by Colombo et al. [490]:

$$\ln K_a = -10.0982 + \frac{4254.05}{T} + 0.2667\ln T \tag{4.12}$$

The reaction mixture is a highly nonideal liquid. The UNIQUAC model is used to calculate the liquid activity coefficients adapting the binary interactions parameters published in [489]. The Soave-Redlich-Kwong (SRK) equation of state is utilized to describe vapor phase behavior. Saturated vapor pressures are calculated via the Antoine equation.

Available raw materials include a pure liquid methanol feed as well as a vapor butenes feed. The feed conditions (i.e., temperature, flowrate, molar composition) are summarized in Table 4.7 taken from Hauan et al. [485].

Two sets of disturbances exist during system operation: (i) a disturbance in the methanol liquid feed flowrate, and (ii) a disturbance in the IB4 inlet composition. The butenes feed flowrate is utilized as a manipulated variable and the MTBE molar composition in the bottom product is treated as the control variable with a desired set point of 98 mol%. Thus the MTBE reactive distillation column is a single input single output (SISO) system.

A nominal MTBE reactive distillation column design is considered as a base case as specified in Table 4.8. The column design is suggested by [491] as the cost-optimal design for the above described MTBE production task via steady-state synthesis with operability considerations. The specific distillation tray design data are taken from Schenk [492].[5]

High fidelity dynamic modelling. The high fidelity model used in this example has been validated against experimental data and has been applied to describe a variety of reactive distillation systems as presented in our previous works [151, 417, 486, 493]. Some key features of this generalized RD model, which enables its prediction accuracy and representation capability for design optimization, are listed as follows:

[5] The number of trays shown in Table 4.8 does not include reboiler or condenser.

Table 4.7 Feed conditions for MTBE production [385]

	Liquid feed	Vapor feed
Temperature (K)	320	350
Flowrate (mol/s)	215.5 ± 10	545
x_{MeOH}	1	0
x_{IB4}	0	0.3578 ± 0.05
x_{NB4}	0	0.6422
x_{MTBE}	0	0
Pressure (atm)	11	11

Table 4.8 MTBE column nominal design

Variable	Value
Number of trays	13
Feed tray locations:	
Liquid methanol feed	Tray 7
Vapor butenes feed	Tray 10
Sieve tray design:	
Diameter	2.5 m
Tray spacing	0.311 m
Active area	5.3 m^2
Weir length	3.66 m
Weir height	0.038 m
Holes area	0.636 m^2
Reactive section	Tray 3–10
Catalyst load	800 kg/tray
Column pressure	8 atm
Reflux ratio	2.75

- A superstructure model formulation to enable the selection of the number of trays and feed tray location via integer variables
- Dynamic material and energy balances for each tray, reboiler, and condenser
- Accurate calculation of non-ideal phase equilibrium via the use of physical property models (in this example, UNIQUAC for liquid activity coefficients and Soave-Redlich-Kwong model for vapor fugacity coefficients)
- Consideration of liquid and vapor, material and energy holdups
- Consideration of liquid hydraulics and liquid level on each tray using modified Francis weir formulation

Table 4.9 Model statistics for the MTBE reactive distillation model [385]

Equations:	
Modeling equations (Eqs. B.1–B.36)	962
Initial conditions (Eqs. B.37–B.38)	52
Variables:	
Algebraic variables	910
Differential variables	52

- Consideration of non-phase equilibrium using Murphree tray efficiencies
- Pressure drop from tray to tray correlated with the vapor flow through the openings at the bottom of each tray and the hydrostatic pressure on each tray
- Detailed calculation of flooding and entrainment correlations and evaluation of minimum allowable column diameter.

The detailed model formulation can be found in Appendix B.2 with relevant nomenclature. The model statistics are presented in Table 4.9. This high fidelity modeling step takes place in PSE's gPROMS® ModelBuilder. Physical properties (e.g., activity/fugacity coefficients, enthalpy, saturated pressure) are calculated using the MultiFlash thermodynamic package which is integrated with gPROMS®.

Model approximation. MATLAB's System Identification/MATLAB® Toolbox is utilized to develop the approximate model for the reactive distillation process, using random input-output sets. For the derivation of the resulting discrete linear state-space model a discretization step, $T_s = 10s$, is selected. The discretized approximate model consists of 6 identified states and the bottom molar fraction of MTBE along with the vapor molar flowrate from the first tray are selected as the measured outputs of the system. The parameters of the state space model are determined as follows.

$$
\begin{gathered}
A = \begin{bmatrix} 0.9898 & 0.0084 & -0.0499 & 0.0024 & 0.0362 & -0.1017 \\ -0.0668 & 0.8165 & 0.0670 & 0.0848 & 0.0179 & -0.0244 \\ 0.1875 & -0.0313 & 0.6780 & -0.1176 & 0.1388 & -1.0977 \\ 0.0229 & 0.4052 & -0.1391 & 0.4649 & -0.4364 & 1.9624 \\ 0.0159 & -0.0877 & 0.1333 & -0.1057 & 0.3208 & 0.4096 \\ -0.0001 & 0.0219 & 0.0457 & -0.0426 & -0.4609 & -0.6961 \end{bmatrix} \\
B = \begin{bmatrix} 0.0005 \\ 0.0058 \\ -0.0023 \\ -0.0336 \\ -0.0098 \\ -0.0025 \end{bmatrix} \qquad C = \begin{bmatrix} 0.0006 & 0.3367 \\ -0.0044 & 4.4020 \\ 0.0096 & -4.4309 \\ 0.0233 & -27.3449 \\ 0.0123 & -11.7250 \\ 0.0114 & -5.7239 \end{bmatrix} \\
D = \begin{bmatrix} 0.3025 & -0.0015 & -0.0079 & 0.0005 & 0.0007 & -0.0002 \\ -83.0519 & 585.6174 & 100.1365 & 57.3670 & 6.1853 & -0.4021 \end{bmatrix}
\end{gathered} \tag{4.13}
$$

Table 4.10 Tuning parameters of the design-dependent explicit MPC of the reactive distillation column [385]

MPC design parameters	Value
OH	2
CH	1
QR	20
R	10^{-7}
$u_{\min}$	490
$u_{\max}$	750
$y_{\min}$	0
$y_{\max}$	1
$d_{\min}$	$\begin{bmatrix}205.5 & 0.3 & 0\end{bmatrix}^T$
$d_{\max}$	$\begin{bmatrix}235.5 & 0.4078 & 1600\end{bmatrix}^T$

Design of the multi-parametric model predictive control. The mpMPC scheme is designed using a similar structure to the binary distillation column example. The design parameters of the controller is presented in Table 4.10.

The MPC problem is translated into a multi-parametric programming problem and the resulting problem is solved in the Parametric OPtimization (POP) Toolbox using the Graph Algorithm [376, 494]. The optimal map of solutions includes 17 critical regions described by the corresponding active sets. Figure 4.5 presents the closed loop validation where the derived controller is tested against the original high fidelity model.

Dynamic optimization. The mixed-integer dynamic optimization problem formulation is introduced to PSE's gPROMS®. The derived design-aware explicit model predictive controllers, which were derived in the previous step are embedded in the optimization formulation. The locally optimal solution of the nonlinear and nonconvex dynamic optimization are presented in Table 4.11.

The operation of the resulting optimal reactive distillation column for 24 h is shown in Fig. 4.6, where feasible operation is maintained in the presence of given disturbances over the horizon of interest.

As it can be observed, the optimal design configuration is able to achieve the operating targets for the bottom purity and flowrate, while minimizing the overall cost. Additionally, the controller is able to guarantee the operability of the reactive distillation column in the presence of disturbances.

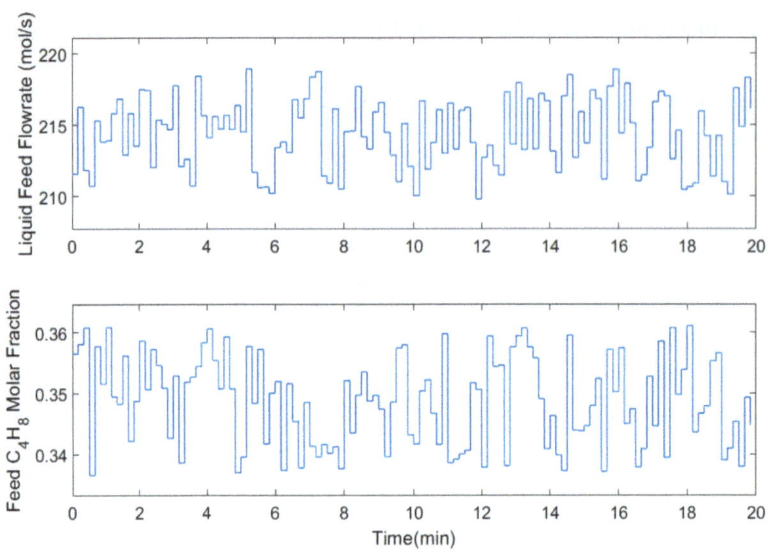

(a) Disturbance profiles.

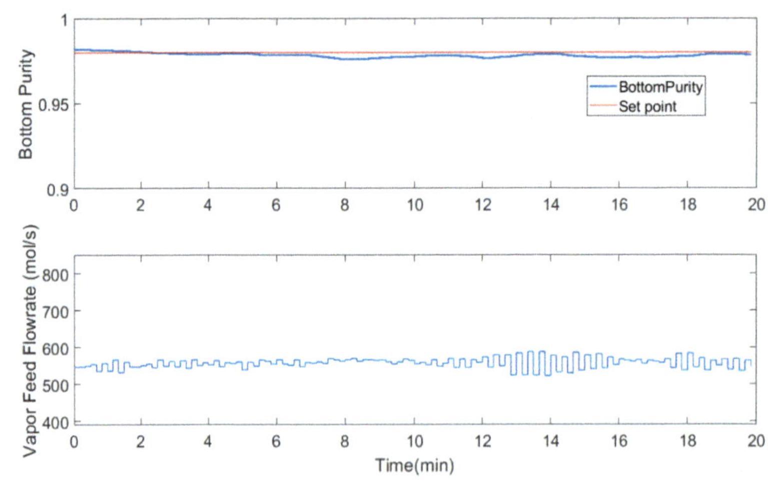

(b) Controller output and input profiles.

Fig. 4.5 Closed-loop validation of the explicit design-dependent controller [385]

Table 4.11 Comparison of the nominal design and closed loop dynamic optimization results for the reactive distillation column example [385]

	Nominal design	Explicit MPC integrated MIDO
Column diameter (m)	2.5	2
Number of trays	13	13
Feed tray location 1	10	10
Feed tray location 2	7	7
Reactive zone interval	(3–10)	(3–10)
Total catalyst mass (kg)	6,400	4,400
Total cost (x 10^6 \$/year)	2.543	2.371

Domestic Cogeneration Unit

The domestic internal combustion engine powered cogeneration example of [495] and Chap. 3 is used here. For the domestic cogeneration unit we assume the possibility of a dual mode of operation, i.e. the unit can either follow an electricity demand (Mode 1 operation) or a hot water demand (Mode 2 operation) based on cost of operation criteria. The decentralized control schemes for this purpose are described in [495] in detail. The unit is assumed to provide heat and power to an area connected to the electricity and natural gas grid, therefore, any power that the unit cannot cover are covered by the grid, at a cost.[6] Similarly surplus of electrical power is provided to the grid, at a revenue and heat in the form of water discarded at a cost. The design aspect is the size of the internal combustion engine. Indicatively, a larger engine can potential cover greater demands but requires a greater investment cost. On the other hand, a smaller engine might be less cost effective in the long term operation of the plant. A sinusoidal demand for electrical power and hot water is assumed. Previously we presented a similar example with Mode 1 operation where the simultaneous design and control employed PI controllers and later a mpMPC was formulated to fit the optimal design [86]. Here, we formulate the explicit controllers to be design dependent, similarly to the three previous examples and proceed with a dynamic design optimization.

High fidelity dynamic modelling. The high 'fidelity model' of the cogeneration unit features the interactions of each component of the unit. It is based on first principles and correlations and is a non-linear DAE system with 379 equations, 15 of which are differential and 6° of freedom. An overview is presented in Tables 4.12 and B.3. The full model can be found in [377].

The necessity of the model approximation and the decentralization of such a system lies with the fact that (i) its complexity would pose a significant challenge for advanced

[6] The cost of electrical power fluctuates between night time and day time which is taken into account in this example.

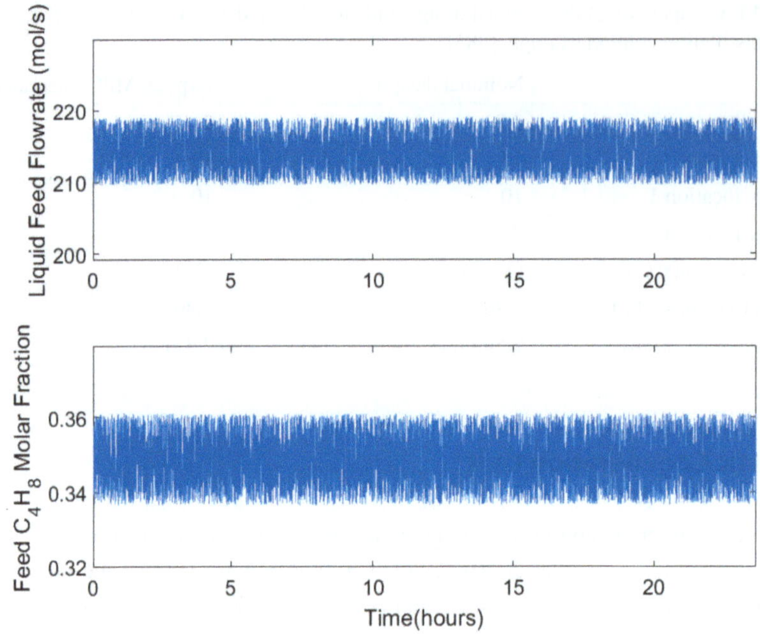

(a) Disturbance profiles.

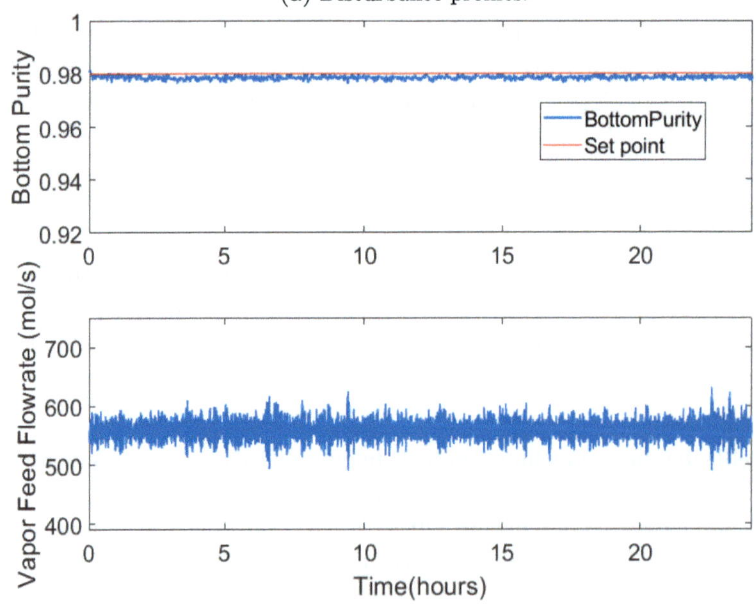

(b) Controller output and input profiles.

Fig. 4.6 Closed-loop profile of the design dependent mpMPC and the integrated process design

Table 4.12 An overview of the residential CHP model [85]

Description	Equation
Throttle valve—Fuel and air manipulation	$\dot{m}_{th} = c_d A_{th} \frac{P_{ab}}{\sqrt{R_\beta T_{ab}}} \psi \left(\frac{P_{ab}}{P_{mn}}\right)$
Manifolds—Pressure difference driven flow for the inlet air and exhaust gases	$\frac{d}{dt} m_{mn} = \dot{m}_{mn,in} - \dot{m}_{mn,out}$ $\dot{V}_{mn,out} = cpf\left(P_{mn} - P_{mn,out}\right)$ $\dot{V}_{mn} \frac{d}{dt} E = \dot{m}_{mn,in} h_{mn,in} - \dot{m}_{mn,out} h_{mn,out}$
Internal combustion engine—Energy and mass balances	$\dot{m}_{ex} = \frac{P_{mn}}{R_\beta T_{mn}} \frac{V_d}{4\pi} \eta_{vl} \omega_{en}$ $\dot{m}_{mn,out}\left(h_{mn,out} + \sum_{i=aircomp.}\left(x_{air,i} h^o_{f,air,i}\right)\right)$ $+\dot{m}_\phi\left(h_\phi + \sum_{j=fuelcomp.}\left(x_{\phi,j} h^o_{f,fuel,j}\right)\right)$ $-\dot{m}_{ex}\left(h_{ex} + \sum_{k=exhaustcomp.}\left(x_{ex,k} h^o_{f,ex,k}\right)\right)$ $- \dot{Q}_f \mid \dot{Q}_{cg\to cw} + \dot{W}_c + \dot{W}_{en}$
Crankshaft—Torque generation	$To_{en} = \frac{P_{meb} V_d}{4\pi}$
Generator—Power generation through torque	$\frac{d}{dt} To_{cl} = \frac{1}{Fl}(To_{en} - To_{cl})$ $Pec = \eta_{en} To_{cl} \omega_{en}$
Engine cooling system—Energy balancces	$\frac{d}{dt} T_i = \frac{\dot{Q}_{in,i} - \dot{Q}_{out,i}}{m_i c_{p,i}}$ $\forall i \in$ engine cooling system components $\dot{Q}_{a\to b} = TC_{ab} A_{ab}(T_a - T_b)$ $\forall a, b \in$ engine cooling system components
Heat exchangers—Energy balances	$Q = UA\Delta T_{mean}$ $\Delta T_{mean} = \frac{\Delta T_{in} - \Delta T_{out}}{log\left(\lvert\frac{\Delta T_{in}}{\Delta T_{out}}\rvert\right)}$.

optimization techniques and (ii) the different modes of operation require advanced control schemes as explained in [495]. The latter can be comprehended by the fact that the process is multi-product, as it is able to produce simultaneously usable heat and power. The dependence between heat and power generation though, makes it impossible to produce both of them simultaneously at the desired level, at all times.

The CHP unit is treated as the interactions of the heat generation subsystem with the power generation subsystem [495]. The power generation subsystem is design dependent as the amount of power generated from the unit is directly affected by the size of the internal combustion engine. The heat recovery subsystem is dependent on the power generation, as it correlates the amount of hot water produced and its temperature to the operating level of the power generation subsystem. Treating the power generation level as a known disturbance (in the case of Mode 1 operation) or as a projected operating level set point (in the case of Mode 2 operation) results only into an indirect design correlation. Therefore, we proceed here considering only the directly affected design dependent power generation subsystem. The formulation for the heat recovery subsystem follows exactly the principles presented in [495].

Model approximation. The approximate model for the power generation subsystems is identified via System Identification in MATLAB®. In order for this to happen, I/O data for a range of different designs are introduced. The input to the system is the opening of the throttle valve which manipulates the amount of air and fuel that enters the combustion chambers of the internal combustion engine. Based on the size of the engine (treated here as a measured disturbance), the power output is affected as shown in Eq. 4.14. The state-space model for the heat recovery subsystem is presented in (4.15).

$$\begin{aligned} x_{T+1} &= 0.9799 x_T + 0.006328 u_{c,T} + 6.516 De \\ y_T &= 7.839 x_T, \quad T_s = 0.1\,s \end{aligned} \tag{4.14}$$

where x_T are the identified states, $u_{c,T}$ is the throttle valve opening, d_T and De is the volume of the internal combustion engine. The output y_T is the electrical power generated via the subsystem.

$$\begin{aligned} x_{T+1} &= \begin{bmatrix} 0.997 & 0.1026 & -0.002958 \\ -0.001527 & 0.9404 & 0.1663 \\ -0.05827 & -0.05636 & 0.179 \end{bmatrix} x_T + \begin{bmatrix} -0.007864 & 0.001107 \\ 0.2801 & -0.03306 \\ -1.28 & 0.1464 \end{bmatrix} u_{c,T} \\ y_T &= \begin{bmatrix} -529.9 & -2.827 & 0.2521 \end{bmatrix} x_T, \quad T_s = 0.1\,s \end{aligned} \tag{4.15}$$

where x_T are the identified states and $u_{c,T}$ are the power generation level and water flow rate in the heat recovery subsystem. The output y_T is the temperature of the hot water at the outlet of the system. Note that depending on the mode of operation of the CHP system one of the inputs is treated as a measured disturbance to the system (i.e. the power generation

Table 4.13 Parameter tuning for the mpMPC of the CHP unit [85]

MPC design parameters	Value
N	3
M	3
$QR_k, \forall k \in \{1, \ldots, \mathrm{N}\}$	31.25
$R1_k, \forall k \in \{1, \ldots, \mathrm{M}\}$	0.1
$x_{\min}$	0
$x_{\max}$	5
$u_{\min}$	0.01
$u_{\max}$	1
$y_{\min}$	0
$y_{\max}$	11
$d_{\min}$	1500
$d_{\max}$	5000
$\Delta u_{\min}$	−0.1
$\Delta u_{\max}$	0.1

level is treated as a measured disturbance in Mode 1 and the water flow rate is treated as a measured disturbance in Mode 2).

Design of the multi-parametric model predictive controller. The multi-parametric model predictive controller problem (based on Eq. 3.1) is formulated and solved with POP [376]. The tuning of the controller is presented in Table 4.13. The tuning of the controllers for the heat recovery subsystem is omitted here as it has been presented previously and is identical to [495].

Note the following:

- The minimum value of the throttle valve opening is not equal to 0. This means that there is a minimum operating level for the CHP unit and that we do not account for the system ability to switch off as part of the control strategy.
- The maximum output is linearly correlated to the size of the internal combustion engine based on the function $y_{\max} = 0.0011De + 4.5148$, where $y_{\max}$ is the maximum power output in $\times 10$ kW and De is the internal combustion engine volume (treated in the context of mpMPC as the disturbance d).

In Mode 1, the control scheme attempts to (i) cover the electrical power demand and (ii) produce water of a predefined temperature, regardless of the flow rate. In Mode 2, the control scheme attempts to produce hot water of (i) a predefined flow rate an (ii) temperature, regardless of the operating level of the power generation subsystem. The most important

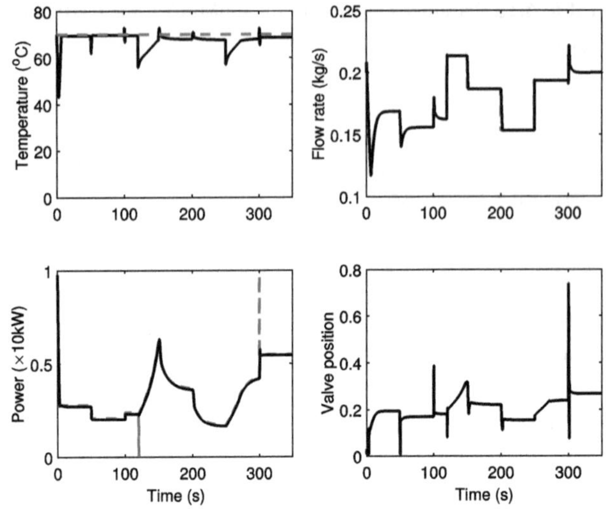

Fig. 4.7 Closed loop validation of the controller against the high fidelity model [85]

limiting factor for the controller performance of the CHP system is to guarantee that there exist no violation of the water temperature above 100°C due to a possible overshoot. Further information regarding this can be found in [495].

Closed-loop validation. Closing the loop of the CHP system involves both operating modes. In Fig. 4.7 the simulation presents a power driven operation for the first 120 s and the last 50 s. A heat recovery driven operation is shown for time between 120 and 300 s. The simulation follows a variable electrical power and hot water flow rate demand. It is assumed that the hot water temperature at the outlet is 70°C. The design of the internal combustion engine for this operation is 1500 cc.

In Figs. 4.8 and 4.9 present the response of the Mode 1 and Mode 2 control scheme respectively for different designs. Note that the same design dependent controller is used for different closed-loop simulations. Despite that, the different designs cause a different response to the system. For example, the throttle valve opening is different between the different designs which leads to different fuel consumption therefore to different operating cost of the CHP unit. Although the difference is relatively small (i.e. the characteristics of the overall profile are similar) those differences suffice to affect the long term operating cost of the unit. The more intense operation of the smaller engine in comparison with the less intense operation of the larger engine is a clear trade-off between investments and operational cost, an aspect that the optimal design takes into account. Table 4.14 shows a snapshot of the explicit actions for an engine size of 1500 and 4500 cc. The upper graphs within Table 4.14 shows the optimal action $u_{c,T}$ applied to the system as a function of the approximate states x_T while the lower graphs the critical region that corresponds to that action. The expressions of the control actions that correspond to the same critical region are identical for different designs. The effect of the different designs to the optimal action is visible from the fact that

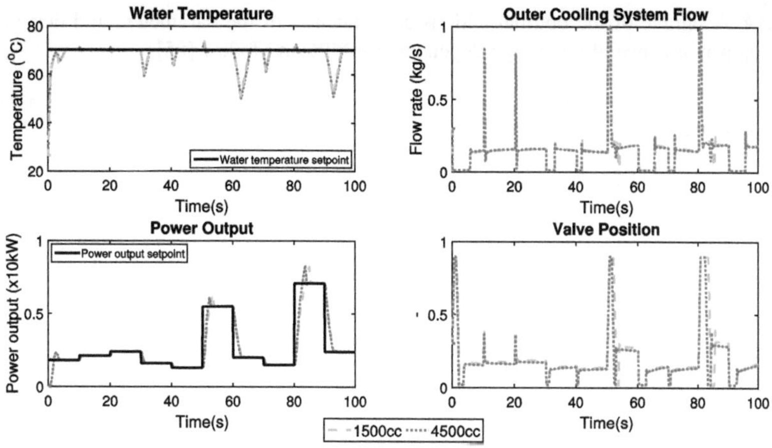

Fig. 4.8 Closed loop performance of the design dependent mpMPC in Mode 1 [85]

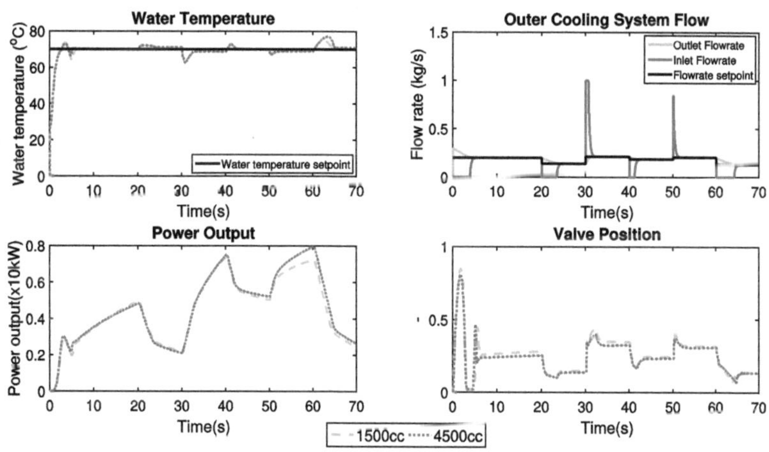

Fig. 4.9 Closed loop performance of the design dependent mpMPC in Mode 2 [85]

for actions corresponding to the same critical region the numerical value that is ultimately applied to the system is different.

Figure 4.10 represents a snapshot of the power generation of the CHP system for different designs under a fixed power demand step change scenario. More specifically, the different responses of the system (y_T) are plotted against the optimal actions ($u_{c,T}$) for three different designs of the internal combustion engine volume (De). This shows that although an advanced design dependent control scheme has been employed there still exists a difference between the closed loop response of the system when $y_T(u_{c,T})$ is regarded.

Dynamic optimization. The dynamic optimization formulation for the CHP system is similar to the previous three examples. More specifically, a sinusoidal demand is introduced for

Table 4.14 A sample of the optimal actions as a function of the parameters and design variable Closed loop performance of the design dependent mpMPC in Mode 1 [85]

$De = 1500$ cc	Time [s]	CR index	Functional form of $u^{1500}_{c,T}$
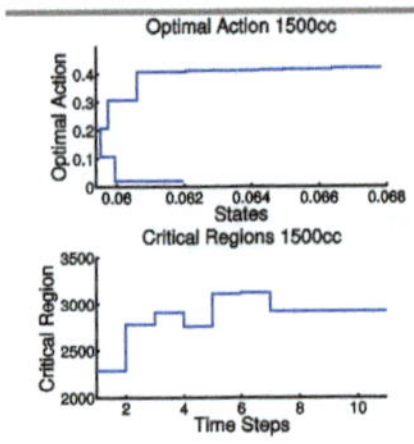	1	2287	0.02
	2	2775	0.02
	3	2915	$3.05x_T + 0.0351u^{1500}_{c,T=2} - 9.8910^{-6}De - 14.81y_T + 14.81y_T^{SP}$
	4	2757	$u^{1500}_{c,T=3} + 0.1$
	5	3112	$u^{1500}_{c,T=4} + 0.1$
	6	3122	$u^{1500}_{c,T=5} + 0.1$
	7	2924	$3.15x_T + 0.004u^{1500}_{c,T=6} - 1.0210^{-5}De - 7.9y_T + 7.9y_T^{SP} + 0.0849$
	8	2924	$3.15x_T + 0.004u^{1500}_{c,T=7} - 1.0210^{-5}De - 7.9y_T + 7.9y_T^{SP} + 0.0849$
	9	2924	$3.15x_T + 0.004u^{1500}_{c,T=8} - 1.0210^{-5}De - 7.9y_T + 7.9y_T^{SP} + 0.0849$
	10	2924	$3.15x_T + 0.004u^{1500}_{c,T=9} - 1.0210^{-5}De - 7.9y_T + 7.9y_T^{SP} + 0.0849$
$De = 4500$ cc	Time [s]	CR index	Functional form of $u^{4500}_{c,T}$
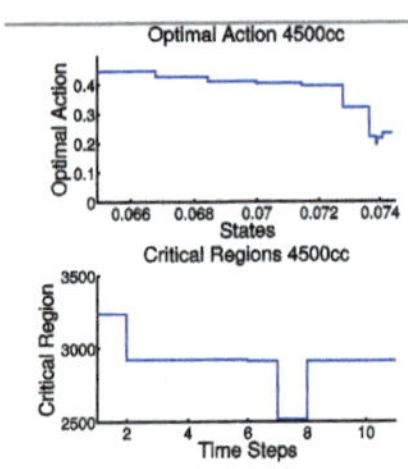	1	$CR_{T=1} = 3239$	$u^{4500}_{c,T=1} = u^{T=0}_{opt} + 0.1$

(continued)

Table 4.14 (continued)

$De = 1500\,\text{cc}$	Time [s]	CR index	Functional form of $u_{c,T}^{1500}$
	2	2924	$3.15x_T + 0.004u_{c,T=1}^{4500} - 1.0210^{-5}De - 7.9y_T + 7.9y_T^{SP} + 0.0849$
	3	2924	$3.15x_T + 0.004u_{c,T=2}^{4500} - 1.0210^{-5}De - 7.9y_T + 7.9y_T^{SP} + 0.0849$
	4	2924	$3.15x_T + 0.004u_{c,T=3}^{4500} - 1.0210^{-5}De - 7.9y_T + 7.9y_T^{SP} + 0.0849$
	5	2924	$3.15x_T + 0.004u_{c,T=4}^{4500} - 1.0210^{-5}De - 7.9y_T + 7.9y_T^{SP} + 0.0849$
	6	2915	$3.04x_T + 0.04u_{c,T=5}^{4500} - 9.8510^{-6}De - 14.81y_T + 14.81y_T^{SP}$
	7	2519	$u_{c,T=6}^{4500} - 0.1$
	8	2915	$3.04x_T + 0.04u_{c,T=7}^{4500}$ $9.8510^{-6}De - 14.81y_T + 14.81y_T^{SP}$
	9	2915	$3.04x_T + 0.04u_{c,T=8}^{4500} - 9.8510^{-6}De - 14.81y_T + 14.81y_T^{SP}$
	10	2915	$3.04x_T + 0.04u_{c,T=9}^{4500} - 9.8510^{-6}De - 14.81y_T + 14.81y_T^{SP}$

both the electrical power and hot water flow rate while the temperature for the hot water is maintained at 70°C. The pricing for the electrical power fluctuates between a high value during the day and a low value during the night time. The purpose of the dynamic optimization problem in this case is dual:

- The determination of the size of the internal combustion engine of the CHP.
- The determination of the operating mode of the system for a given demand scenario and pricing scheme.

The objective function is formulated as a Total cost function that includes (i) the investment cost of the CHP, (ii) the operating cost of the unit as a function of the amount of fuel

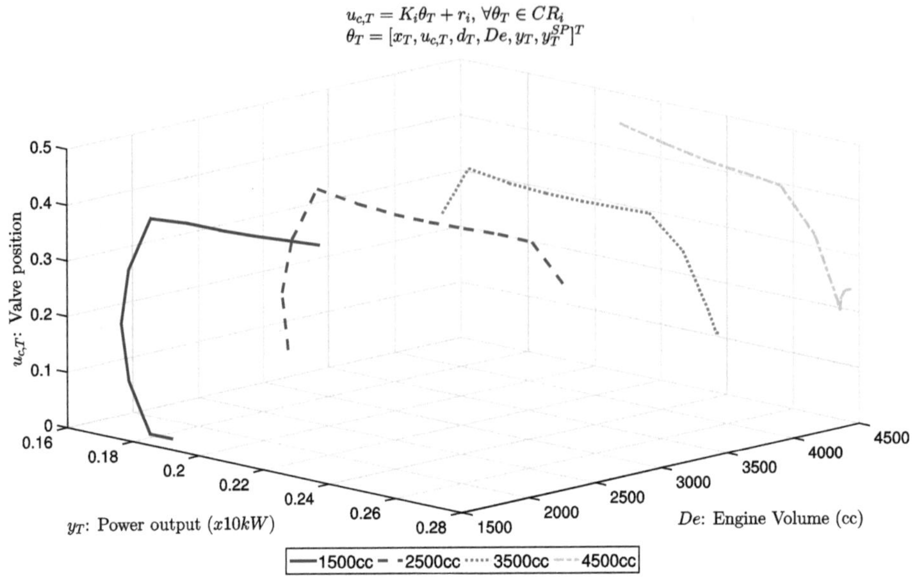

Fig. 4.10 State and optimal control action profiles with respect to different design configurations Closed loop performance of the design dependent mpMPC in Mode 1 [85]

required for electrical power and heat production, (iii) the cost of external resources required to cover possible demand shortages, (iv) the cost of discarding surplus heat (in the form of hot water) and (v) the revenues from selling surplus power to the grid.

The MIDO algorithm converges to the an operation based on the day and night pricing of electrical power, i.e. the system prefers to operate on a power production driven operation during the day when the electricity price is higher and revert to a heat generation driven demand during night time. Furthermore, the size of the internal combustion engine converges to 2203 cc, a value that does not correspond to the upper or lower bound of the design variable. It is clear through testing several demand scenarios (and from previous studies [86]) that the operating mode of the system is primarily a matter of fuel and electricity price. More specifically, the choice for the operating mode of the CHP system, although here optimized for a given demand scenario and electricity price scheme, is a matter that needs to be determined as the fluctuation in demand and pricing happens. Therefore, this is a matter of an economic/operational optimization for demand side uncertainty, such as a scheduling formulation as described in [147].

Table 4.15 Computational time statistics

Case study	Time scale	
	(MI)DO problem	mpMPC offline problems
Simple tank	<1 min	<5 min
CSTR	<10 min	<5 min
Distillation column	<2 h	<10 min
CHP	<12 h	<1 h (for all problems combined)

Computational Statistics and Requirements

All computational experiments were performed on a computer running Windows 7 Enterprise, SP1, MATLABR2015b® 64bit, gPROMS®4.2 64bit and Visual Studio 2012. The computer was equipped with 16GB of RAM and Intel®Core™i7-4790 clocked at 3.60 GHz.

The computational time scales for the mpMPC and MIDO problems are reported in Table 4.15

4.3 Concluding Remarks

We presented a framework for the application of design and control optimization via multi-parametric programming though four case studies. We demonstrated the procedure to develop a design-dependent mpMPC and its integration into an MIDO for design optimization in the CSTR example. The binary distillation column example was revisited and the size and key tray locations were determined via a design MIDO formulation and mpMPC with the latter taking into account discrete alternatives. The reactive distillation column demonstrated the applicability of the framework on intensified processes where the range of operability is limited. The common basis for these three examples is that the operation of the aforementioned equipment focuses on a simple objective, which is to minimize the capital and operating costs while delivering certain product quality metrics.

On the contrary to the previous three examples the case of the CHP is different in the sense that its operation is variable, i.e. the system should be able to follow different operating set points at different times. Although a simultaneous design and control approach provides great insight regarding its operation, particularly the change between operating modes within the day, it is not able to determine the long term operation of the system efficiently. In other words, a decision about a long term characteristic (such as the size of the internal combustion engine) of a variably operating system (alternating operating set points) cannot be made without considering the optimality of the aforementioned set points, especially in a case of a multi-product process. Therefore, subject of our current work is the simultaneous consideration of the optimality of the set points based on mid term financial criteria, i.e. the inclusion of simultaneous reactive scheduling formulations of such systems.

5 Process Scheduling and Control via Multi-parametric Programming

In this chapter, we extend the PAROC framework introduced in Chap. 3 to integrate process scheduling with the mpMPC strategy. We develop offline maps of optimal scheduling actions accounting for the closed-loop dynamics of the process through a surrogate model formulation that incorporates the inherent behavior of the control scheme. The surrogate model is designed to translate the long term scheduling decisions to time varying set points and operating modes in the time scale of the controller. The continuous and binary scheduling decisions are explicitly taken into account in the multi-parametric model predictive controllers. To showcase the framework, we consider the illustrative CSTR example discussed in Chap. 3 and use it as a multiproduct reactor to produce three products. We then extend the problem to account for two reactors operating in parallel [95].[1]

5.1 Overview

The traditional approach to assess the multi-scale operational activities sequentially often leads to suboptimal solutions as each problem dictates different, and sometimes conflicting, objectives. The most recent advances in the field of operational research and the rapid reduction in the cost of computer hardware have enabled the integration of multi-scale decision making mechanisms [147, 496, 497]. Production scheduling and process control are two layers in the process operations that are highly dependent due to the volume of reciprocal information flow. Typically, the process schedule coordinates the production sequence, production times, and inventory levels based on the market dynamics. Process control, on

[1] Reprinted with permission from B. Burnak, J. Katz, N. A. Diangelakis, and E. N. Pistikopoulos, "Simultaneous process scheduling and control: A multiparametric programming-based approach", Industrial & Engineering Chemistry Research, vol. 57, no. 11, pp. 3963–3976, 2018. Copyright (2018) American Chemical Society.

B. Burnak et al., *Integrated Process Design and Operational Optimization via Multi-parametric Programming*, Synthesis Lectures on Engineering, Science, and Technology, https://doi.org/10.1007/978-3-032-10181-5_5

the other hand, delivers the production targets with the existence of operational uncertainty, measured/unmeasured process disturbances, and plant-model mismatch. These layers are typically addressed independently and sequentially due to the hierarchical nature of the underlying problems. The isolation between the decisions from different layers can result in suboptimal, or even infeasible operations [88, 181].

Individual assessment of the scheduling and control problems requires some assumptions that neglect the dynamics introduced by their complements. The scheduling problem utilizes static tables comprising the process time constants for the transitions between the operating modes of the system. These time constants are typically obtained by exhaustive closed-loop simulations conducted offline. Consequently, the static tables fail to represent the closed-loop dynamics of the system due to the lack of an underlying high-fidelity model [90, 162].

A simultaneous approach for process scheduling and control reconstructs the two problems as a unified problem. The reformulated problem takes into account the degrees of freedom of the two subproblems simultaneously, leading to an augmented feasible space. This allows the chemical plant to respond to rapidly changing market conditions while maintaining feasible and profitable operation. These changes include but are not limited to the market demand, price, and the spectrum and specifications regarding the products manufactured in the chemical plant. Furthermore, fluctuating operating costs require flexibility in the process scheduling [181]. Therefore, a chemical process needs integrated decisions that enable higher adaptability and operability to remain competitive in the market [498]. There have been some attempts over the years to tackle the two aspects of operational optimization in an integrated framework. An indicative list of these contributions is presented in Table 5.1.

Over two decades of academic literature on integrated approaches for the process scheduling and control problem has focused on a systematic methodology to overcome the following fundamental challenges [99]:

(i) *Discrepancies in objectives*: The schedule and control formulations are designed to deliver specialized tasks in a process. The former aims to bring profitable operation by taking into account the operational aspects such as the process economics, raw material and equipment availability, and product specifications; while the latter involves real-time manipulation of select process variables to meet the targeted product specifications. These scheduling and control goals are not always aligned and frequently require the compromise of one of their respective objectives.
(ii) *Discrepancies in time-scales*: A typical control horizon varies between seconds and minutes, whereas the scheduling horizon is on the order of hours or weeks. Therefore, integration of the two distinct problems into a unified formulation creates a large scale, stiff system due to the order of magnitude differences in their respective time scales [99, 162]. Following direct solution approaches for the reformulated unified problem have been shown to be computationally intractable [109].

Table 5.1 An indicative list of process scheduling and control in literature [95]

Author (year)	Contribution
Grossmann and co-workers (2006a, 2006b, 2007, 2010, 2011, 2012, 2014) [90, 132, 184–186, 499, 500], Gudi and co-workers (2010) [501], Biegler and co-workers (2012, 2015) [91, 109], You and co-workers (2013) [92], Daoutidis and co-workers (2020) [502], Flores-Tlacuahuac and co-workers (2020) [503]	Simultaneous/Decomposition (MI)DO or (MI)NLP and open-loop optimal control
Pistikopoulos and co-workers (2003a, 2003b) [183, 504], You and co-workers (2012) [505]	Simultaneous/Decomposition (MI)DO schedule and P-PI-PID control
Allcock and co-workers (2002) [182], Espuña and co-workers (2013) [506], Baldea and co-workers (2014, 2015) [162, 507]	Simultaneous/Decomposition algorithms using control/dynamics aware scheduling models
Biegler and co-workers (1996) [179], Barton and co-workers (1999) [259], Nyström and co-workers (2005) [508], Marquardt and co-workers (2008) [89], Ierapetritou and co-workers (2012) [153], You and co-workers (2013) [92]	Simultaneous/Decomposition algorithms via (MI)DO reformulation to (MI)NLP
Puigjaner and co-workers (1995) [172], Pistikopoulos and co-workers (2013, 2014, 2016) [147, 152, 509], Rawlings and co-workers (2012, 2013) [125, 128], Swartz and co-workers (2020) [101], Rawlings, Maravelias and co-workers (2019) [129], Ierapetritou and co-workers (2020) [510], Ricardez-Sandoval and co-workers (2021) [511]	Control theory in scheduling problems
Marquardt and co-workers (2011) [192], Pistikopoulos and co-workers (2016, 2017, 2017) [147, 175, 512], Ierapetritou and co-workers (2016) [166], Baldea and co-workers (2020) [96], Ierapetritou and co-workers (2020) [161]	Advanced control and (MI)NLP scheduling schemes
Reklaitis and co-workers (1999) [513], Floudas and co-workers (2004, 2007) [514, 515], Ricardez-Sandoval and co-workers (2015, 2017) [4, 5], You and co-workers (2015) [516], Ricardez-Sandoval and co-workers (2021) [167], Ricardez-Sandoval and co-workers (2021) [517]	Scheduling under uncertainty
Reklaitis and co-workers (1996) [518], Floudas and co-workers (2004) [519], Grossmann (2005) [496], Nyström and co-workers (2009) [498], Engell and Harjunkoski (2012) [88], Baldea and co-workers (2014) [181], You and co-workers (2015) [516], Ierapetritou and co-workers (2016) [166], Baldea, Mitsos and and and and co-workers (2020) [520]	Review articles on scheduling and control and methodologies

In this study, we propose a surrogate model formulation that bridges the inherent gap between the schedule and control formulations. The surrogate model is designed to translate the fast closed-loop dynamics to the slower scheduling dynamics, while providing corrective time varying targets for the controller. We utilize the reactive scheduling approach introduced by Subramanian et al. [128], and adapted in a multi-parametric framework by Kopanos and Pistikopoulos [152], formulating a state-space representation that is implemented in a rolling horizon framework. This formulation is solved once and offline via multi-parametric programming techniques, deriving optimal scheduling decisions as affine functions of the product demand scenarios. The derivation of the controllers, on the other hand, is adapted from the PARametric Optimisation and Control (PAROC) framework [227], which provides a systematic methodology to design advanced model-based controllers via multi-parametric programming. The framework is tailored to account for the scheduling decisions explicitly.

The remainder of the chapter is organized as follows. The problem is first defined in Sect. 5.2. The proposed methodology to derive integrated process schedule and control is described in Sect. 5.3 in detail and showcased on a three product CSTR, and two CSTRs in parallel in Sect. 5.4. The concluding remarks on the application of the framework and the future directions to extend its implementation is provided in Sect. 5.5.

5.2 Problem Definition

The following generalized problem definition presents the long term (schedule) and short term (control) objectives simultaneously.

(i) *Given*: A dynamic model of the system of interest, process constraints regarding the safety issues and product specifications, unit cost for inventory, a scenario of the market conditions.
(ii) *Determine*: Production sequence, target production rate, optimal control actions to achieve the target production rate.
(iii) *Objective*: Minimize the total cost comprising the inventory, transition, and raw material costs.

We propose a systematic framework that features the following:

- A single high fidelity model based on which we seek to derive an integrated scheduling and control scheme,
- A scheduling scheme, aware of the short term dynamics,
- A control scheme, aware of the longer term scheduling/operational decisions,
- An offline map of optimal short term and long term operational actions.

5.3 Problem Formulation

The problem defined in Sect. 5.2 is represented with a generalized mathematical model in the form of a MIDO problem, detailed in Chap. 2 and given by Eq. 2.1.

In this chapter, we use of the PAROC framework to decompose the overall problem into two main steps. The proposed methodology consists of (i) acquiring scheduling dependent explicit control schemes, and (ii) developing long term scheduling strategies based on the high fidelity process model featuring the control scheme derived in the first step.

5.3.1 Incorporating Process Scheduling in the PAROC Framework

The first step of the proposed method is to acquire a mathematical high-fidelity model to describe the system of interest with sufficient accuracy, similar to the design and control problem discussed in Chap. 4. These models are typically very large in size and/or complex in nature, rendering difficult to apply an advanced optimization algorithm. Therefore, the original mathematical model is approximated or reduced in size via the existing algorithms in the literature. Note that the approximate model features the scheduling aspects of the system as additional dimensions in order to generate a schedule-aware control scheme. A model predictive control (MPC) scheme is constructed using the approximate model, and solved multi-parametrically (mpMPC) to generate offline maps of optimal control actions. These maps are embedded into the original mathematical model, and a control-aware approximate model is derived to describe the closed-loop behavior of the system. The resulting model is used to derive offline maps of (i) long term decisions regarding the operational feasibility and profitability and (ii) a surrogate model to bridge the gap between the short term and long term decisions. The offline maps are validated against the high-fidelity model used in the first step. A schematic representation of the proposed methodology is presented in Fig. 5.1. The closed-loop implementation of the framework and the fundamental interactions between different layers of models for the integration of schedule and control are depicted in Fig. 5.2.

Note the following advantages of solving the multi-parametric counterparts of the schedule, control, and time scale bridging surrogate model:

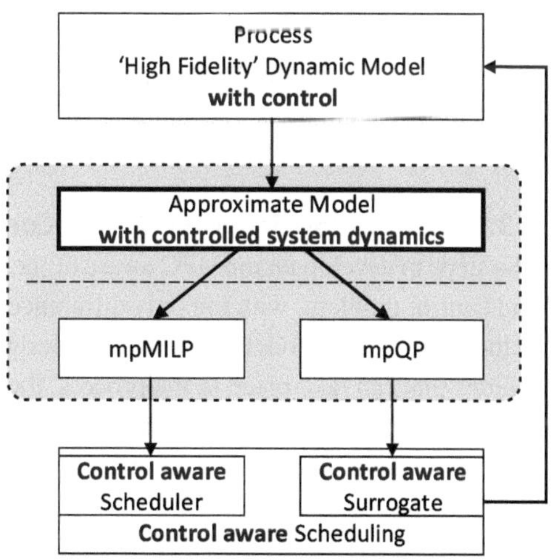

Fig. 5.1 Proposed methodology to derive integrated schedule and control (adapted from [95, 512]). Actions within the gray area happen once and offline

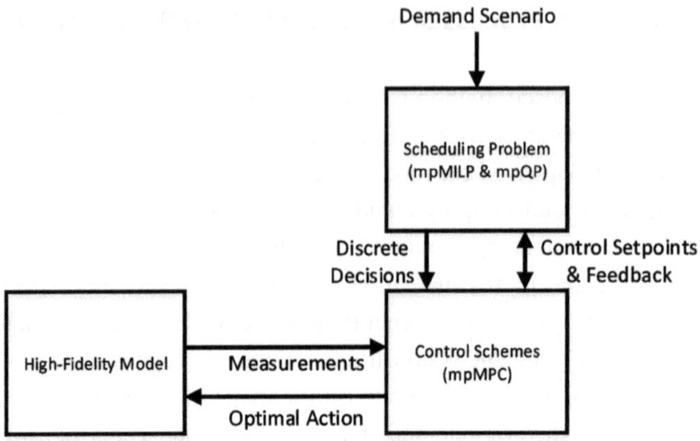

Fig. 5.2 Interactions between the decision making layers for the scheduling and control problems [95]

- Offline maps of optimal operations at both long and short terms are acquired as explicit expressions.
- Online computational time for the optimal control problem is reduced to a simple look-up table algorithm and evaluation of an affine function. Such significant reduction enables the application of the framework to system with fast dynamics.
- The offline maps of solutions can allow for the integration of the design of the process/equipment with the schedule and control in a dynamic optimization framework, which will be discussed extensively in Chap. 6.

Following are the fundamental steps of PAROC in further detail, tailored specific to the needs of the simultaneous scheduling and control problem.

5.3.1.1 Designing Schedule-Aware Controller

The steps to develop an mpMPC aware of the scheduling decisions are similar to the design and control problem, with the only difference that the scheduling decisions are represented in the approximate model rather than the design variables. To account for different operating regions that can take place in the process, the approximate model is comprised of multiple state space systems that are connected through logical disjunctions in the controller and the scheduler. Therefore, we first present the multiple state space models in the following form.

$$\begin{aligned} x^q_{t_c+1} &= A^q x^q_{t_c} + B^q u_{t_c} + C^q [d^T_{t_c}, Sc^T_{t_c}]^T \\ \hat{y}_{t_c} &= D^q x^q_{t_c} + E^q u_{t_c} + F^q [d^T_{t_c}, Sc^T_{t_c}]^T \end{aligned} \tag{5.1}$$

where index t_c indicates the discrete time steps sampled in control scale, superscript q denotes the index of the linear model, and $\hat{y}$ is the output predicted by the approximate model. Note that the states $x_{t_c}^q$ can be concatenated into a single vector, x_{t_c}.

Equation 5.1 is a generalized reduced expression that represents the dynamics between the manipulated inputs, measured disturbances, and outputs of the system. Note that the reduced model also features the degrees of freedom of the system that may be fixed by the scheduler and unavailable to the controller, Sc_{t_c}.

The developed model is used in the mpMPC formulation given by Eq. 3.1 to derive the explicit optimal control law that can be used for the entirety of different operating regions, as presented by Eq. 5.2.

$$\begin{aligned} & u_j(\theta) = K_n\theta + r_n, \forall\theta \in CR_n \\ & \theta := \left[x_{t_c=0}^T, u_{t_c=-1}^T, d_{t_c=0}^T, Sc_{t_c}^T, (y_{t_c}^{SP})^T, (u_{t_c}^{SP})^T, Y_{t_c}^T, y_{t_c=0}^T\right]^T \\ & CR_n := \{\theta \in \Theta \mid L_n\theta \leq b_n\}, \forall n \in \{1, 2, \ldots NC\} \\ & \forall j \in \{0, 1, \ldots, M_c\}, \forall t_c \in \{0, 1, \ldots, N_c\} \end{aligned} \tag{5.2}$$

where θ is the set of uncertain parameters measured at $t_c = 0$, u_{t_c-1} is the optimal control action at the previous time step, CR_n is the active polyhedral partition of the feasible parameter space, NC is the number of critical regions CR_n, and Θ is defined as a closed and bounded set. Note that inclusion of scheduling level decisions, i.e. $y_{t_c}^{SP}$, $u_{t_c}^{SP}$, and Y_{t_c} in the parameter space enables mpMPC to account for any future changes in the operational level a priori within the range of the output horizon. The developed explicit optimal control law is then validated against the high fidelity model at the entirety of the operating range.

5.3.1.2 Designing Control-Aware Scheduler

Production scheduling of a chemical process formulated as a general MILP problem can also be represented by a state space model [125, 128]. Multi-parametric counterpart of this class of reactive scheduling problems and its solution is described extensively in Kopanos and Pistikopoulos [152]. This approach yields an optimal map of solutions under potential disruptions in the course of operation prior to the occurrence of the event. The explicit form of the schedule significantly reduces the computational cost of repetitive evaluations after every disruptive event. However, the sampling time of the state space model is typically too large to account for the dynamic considerations inherent to the process. Hence, such an approach suggests utilization of static transition tables based on exhaustive testing [162] that create plant-model mismatch since they are agnostic to the real system dynamics.

In this section, we introduce a two level scheduling scheme with a hierarchical order: (i) upper level schedule for the regulation of the economic considerations and operational feasibility based on the formulation of Kopanos and Pistikopoulos [152], and (ii) lower level surrogate model to bridge the time scales between the control and the upper level schedule based on the closed loop behavior of the high fidelity model. The surrogate model further aims to remedy the plant-model mismatch introduced by the schedule.

Step 1: High-fidelity model with controller embedded. The control scheme derived in the earlier phase (Eq. 5.2) is embedded in the original high fidelity model.

Step 2: Approximate models. A discrete time state space model is derived based on the closed loop behavior of the high fidelity model. The input-output relationship focuses on capturing the overall response of the closed loop system to the step changes in the output set points and input reference points. Note that the discretization time of the identified model for the upper level scheduler is several orders of magnitude larger than the mpMPC. Therefore, we introduce a surrogate model formulation to translate the upper level scheduling decisions in the first scheduling time step into the control time steps. This translation is carried out by resampling the identified scheduling model with a discretization step matching the output horizon of the mpMPC. The resampled model is used as the governing constraint in the surrogate model formulation, as described in detail in the next step.

Step 3: Design of the multi-parametric schedule and surrogate model. The multi-parametric schedule is formulated with an objective to account for the economic considerations and operational feasibility, subjected to the corresponding approximate model derived earlier, as described in Kopanos and Pistikopoulos [152]. The resulting formulation creates a mpMILP that treats the disruptive scheduling events as parameters described in Eq. 5.3.

$$
\begin{aligned}
\min_{\tilde{u}_{t_s}, Y_{t_s}} J(\theta) = & \sum_{t_s=1}^{N_s} \alpha^T \tilde{x}_{t_s} + \sum_{t_s=0}^{N_s-1} \beta^T \tilde{tr}_{t_s} + \sum_{t_s=0}^{N_s} \phi^T \tilde{u}_{t_s} \\
s.t. \quad & \tilde{x}_{t_s+1} = A_1 \tilde{x}_{t_s} + B_1 \tilde{u}_{t_s} + C_1 \tilde{d}_{t_s} \\
& \tilde{tr}_{t_s} = A_2(\tilde{x}_{t_s} - \tilde{x}_{t_s-1}) + B_2(\tilde{u}_{t_s} - \tilde{u}_{t_s-1}) \\
& \tilde{u}_{t_s} = \left[\tilde{Sc}_{t_s}^T, (\tilde{y}_{t_s}^{SP})^T, (\tilde{u}_{t_s}^{SP})^T\right]^T \\
& \theta = \left[\tilde{x}_{t_s=0}^T, \tilde{x}_{t_s=-1}^T, \tilde{u}_{t_s=-1}^T, \tilde{d}_{t_s}^T\right]^T \\
& \tilde{x}_{\min,t_s} \le \tilde{x}_{t_s} \le \tilde{x}_{\max,t_s} \\
& \tilde{tr}_{\min,t_s} \le \tilde{tr}_{t_s} \le \tilde{tr}_{\max,t_s} \\
& \tilde{u}_{\min,t_s} Y_{t_s} \le \tilde{u}_{t_s} \le \tilde{u}_{\max,t_s} Y_{t_s} \\
& \forall t_s \in \{0, 1, \ldots, N_s\}
\end{aligned}
\tag{5.3}
$$

where the tilde ($\sim$) sign denotes a scheduling level counterpart of the variable, $\tilde{x}$ is the operational level and the inventory, $\tilde{tr}$ denotes transition to a different operational mode, and the Greek letters α, β, and ϕ are the corresponding cost parameters. Note that additional constraints can be included in Eq. 5.3 regarding the needs of the specific problem. The linear state space matrices represent the closed loop dynamics of the system, and acquired through the MATLAB® System Identification Toolbox™ or a model reduction technique, as described by Diangelakis [521]. The multi-parametric solution of Eq. 5.3 provides explicit affine expressions of the optimal scheduling actions as functions of the system parameters, as defined in Eq. 5.4.

$$\begin{aligned}
&[\tilde{u}_{t_s}^T(\theta), Y_{t_s}^T(\theta)]^T = \tilde{K}_n\theta + \tilde{r}_n, \forall\theta \in CR_n \\
&\theta := \left[\tilde{x}_{t_s=0}^T, \tilde{x}_{t_s=-1}^T, \tilde{u}_{t_s=-1}^T, \tilde{d}_{t_s}^T\right]^T \\
&CR_n := \{\theta \in \Theta \mid \tilde{L}_n\theta \leq \tilde{b}_n\}, \forall n \in \{1, 2, \ldots NC\} \\
&\forall t_s \in \{0, 1, \ldots, N_s\}
\end{aligned} \tag{5.4}$$

Due to approximation of the scheduling model and the large discretization time, there exists a plant-model mismatch that is handled by a surrogate model formulated as a mp(MI)QP. Therefore, we utilize the formulation presented in Eq. 5.5 to minimize the aforementioned mismatch.

$$\begin{aligned}
\min_{S_{t_{sm}}} \quad & J(\theta) = \sum_{t_{sm}=0}^{N_{sm}} \left(S_{t_{sm}} - \tilde{u}_{t_s}\right)^T R' \left(S_{t_{sm}} - \tilde{u}_{t_s}\right) \\
s.t. \quad & x'_{t_{sm}+1} = Ax'_{t_{sm}} + BS_{t_{sm}} \\
& y_{t_{sm}} = Cx'_{t_{sm}} + DS_{t_{sm}} \\
& S_{t_{sm}} = \left[Sc_{t_{sm}}^T, (y_{t_{sm}}^{SP})^T, (u_{t_{sm}}^{SP})^T\right]^T \\
& \theta = \left[\tilde{u}_{t_{sm}}^T, Y_{t_{sm}}^T\right] \\
& x'_{\min,t_{sm}} \leq x_{t_{sm}} \leq x'_{\max,t_{sm}} \\
& y_{\min,t_{sm}} \leq y_{t_{sm}} \leq y_{\max,t_{sm}} \\
& S_{\min,t_{sm}} Y_{t_{sm}} \leq S_{t_{sm}} \leq S_{\max,t_{sm}} Y_{t_{sm}} \\
& \forall t_{sm} \subset \{0, 1, \ldots, N_{sm}\}
\end{aligned} \tag{5.5}$$

Equation 5.5 poses a mpQP problem that reinterprets the scheduling actions $\tilde{u}_{t_s}$ in the time steps of the controller. $Sc_{t_{sm}}$ is directly passed to the process, and the set points $y_{t_{sm}}^{SP}$ and $u_{t_{sm}}^{SP}$ are determined to be used by the controller. Δt_{sm} is based on the output horizon of the mpMPC ($\Delta t_{sm} = \Delta t_c N_c$), and N_{sm} is selected such that the surrogate model horizon can account for the first scheduling time step in its entirety (i.e. $N_{sm} \geq \Delta t_s / \Delta t_{sm}$). The multi-parametric solution to Eq. 5.5 yields an offline map of optimal scheduling actions and set points for the controller, allowing for fast reevaluation of the scheduling decisions under varying market conditions. The surrogate model formulation utilizes a linear state space representation of the closed loop dynamics of the system. Therefore, the number of state space models required to capture the complete dynamics is dependent on the complexity of the high fidelity model and the size of the explicit control law. The validity of these surrogate models representations is assured in the subsequent step.

Note that binary decisions $Y_{t_{sm}}$ from Eq. 5.3 are treated as continuous uncertain parameters. Oberdieck et al. [483] presents a rigorous proof through Basic Sensitivity Theorem that relaxation of the binary parameters yields the exact solution in this class of problems. Equivalently, one can generate 2^n mpQP problems to exhaustively enumerate all combinations of binary parameter realizations, where n is the number of binary parameters.

Step 4: Closed-loop validation. Overall validation of the integrated schedule-control scheme is performed in a rolling horizon fashion through utilizing the maps of solutions generated with Eqs. 5.2, 5.4, and 5.5 simultaneously on the high fidelity model. The overall system is subjected to randomized market conditions that is updated in the time steps of the scheduler to yield the input and output trajectories in the scheduling and control levels.

Note that the framework assumes an update in the disruptive events at the time steps of the schedule. Any further scheduling level disturbances in between these time steps can be addressed by reevaluating the set points through the surrogate model to remedy a potential performance degradation. The process disturbances, on the other hand, are accounted for by the controller of which dynamics are embedded in the scheduler.

The following section presents the application of the framework on (i) a CSTR with three reactants and three outputs, and (ii) two CSTRs operated in parallel.

5.4 Examples

5.4.1 Example 1 Single CSTR with Three Inputs and Three Outputs

The CSTR example explored in Chaps. 3 and 4 has a single product to be manufactured throughout the lifespan of the unit. However, processes often have multiple products to be manufactured from a single line to reduce the capital investments. Therefore in this example, we extend the illustrative example covered in earlier chapters to account for the scheduling aspects in a multiproduct task.

Problem definition. This case study, adapted from Flores-Tlacuahuac and Grossmann [90], considers an isothermal CSTR designed to manufacture three products on a single production line, as depicted in Fig. 5.3. In the figure, R_i denotes the ith reactant, P_j denotes the jth product, and $Demand_{P_j}$ denotes the demand rate for product P_j. The problem statement encompasses the following:

(i) *Given*: A high-fidelity model of the three product CSTR, unit cost for inventory, a scenario of product demands.
(ii) *Determine*: Production sequence, production rates, optimal control actions to achieve the target production rate and to reach the threshold purity.
(iii) *Objective*: Minimize the total cost comprising the inventory and transition costs.

Based on the described problem definition, the control scheme aims to determine the optimal transitions between the production periods of three products through tracking a time variant product concentration set point to maintain a certain level of purity threshold. The

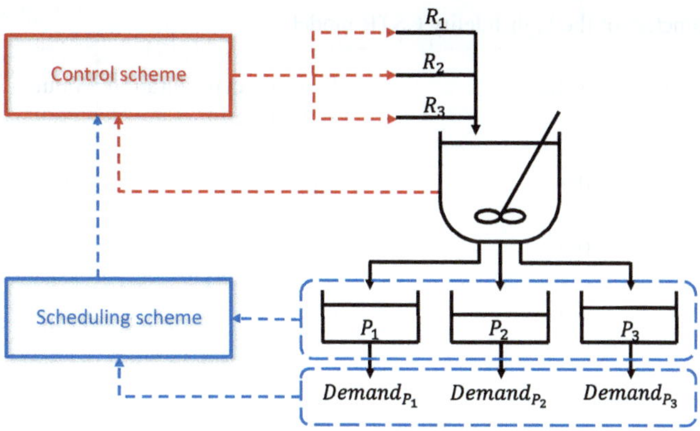

Fig. 5.3 CSTR flowsheet with the control and scheduling layers [95]

controller is designed to deliver this short term objective by manipulating the feed composition at the inlet of the reactor, and monitoring the states of the system. To obtain the longer term objectives, we utilize a scheduling scheme to minimize the operating and inventory costs, while satisfying a continuous demand rate for each product. The scheduler aims to determine the optimal production sequence and manufacturing time, while accounting for the inventory levels in the storage tanks and a demand scenario. The scheduling decisions are passed on to the controller as set points and operating modes.

Note that different from Flores-Tlacuahuac and Grossmann [90], this work relaxes the assumption of constant product demand rate profile, and considers a variable demand rate profile.

High-fidelity dynamic model. Three irreversible reactions take place in parallel in the CSTR reaction network given in Eq. 5.6.

$$\begin{aligned} 2R_1 &\xrightarrow{k_1} P_1 \\ R_1 + R_2 &\xrightarrow{k_2} P_2 \\ R_1 + R_3 &\xrightarrow{k_3} P_3 \end{aligned} \tag{5.6}$$

where k_1, k_2, and k_3 are the rate constants of the respective reactions. Note that production of P_1 requires only R_1, which also features as one of the raw materials of products P_2 and P_3. Hence, the given reaction network yields P_1 as a by-product during the production phases of P_2 and P_3. The by-product concentration degrades the purity of the product of interest, and needs to be accounted for by the control scheme to achieve high selectivity.

The high-fidelity model that describes the dynamic behavior of the CSTR comprises mole balances (Eq. 5.7) and power law kinetic expressions for elementary reactions (Eq. 5.8).

Table 5.2 Parameters of the high-fidelity CSTR model

Reaction rate constants	Value	Reactant concentration at the feed	Value
k_1	0.1	$C^f_{R_1}$	1.0
k_2	0.9	$C^f_{R_2}$	0.8
k_3	1.5	$C^f_{R_3}$	1.0

$$\begin{aligned}\frac{dC_{R_i}}{dt} &= \frac{Q_{R_i} C^f_{R_i} - Q_{total} C_{R_i}}{V} + \mathcal{R}_{R_i}\\ \frac{dC_{P_j}}{dt} &= \frac{Q_{total}(C_{P_j} - C_{P_j})}{V} + \mathcal{R}_{P_j}\end{aligned} \tag{5.7}$$

$$\begin{aligned}\mathcal{R}_{R_1} &= -2\mathcal{R}_{P_1} - \mathcal{R}_{P_2} - \mathcal{R}_{P_3}\\ \mathcal{R}_{R_2} &= -\mathcal{R}_{P_2}\\ \mathcal{R}_{R_3} &= -\mathcal{R}_{P_3}\\ \mathcal{R}_{P_1} &= k_1 C^2_{R_1}\\ \mathcal{R}_{P_2} &= k_2 C_{R_1} C_{R_2}\\ \mathcal{R}_{P_3} &= k_3 C_{R_1} C_{R_3}\end{aligned} \tag{5.8}$$

where C denotes the concentration, Q is the volumetric flow rate, V is the volume of the CSTR, $\mathcal{R}$ is the reaction rate, superscript f denotes the feed to the CSTR, R_i and P_j are the indices for the ith reactant and j^{th} product, respectively. The system parameters are given in Table 5.2.

The total volumetric flow rate is defined as the sum of reactant flow rates at the inlet of the reactor. Note that constant volume reactor is assumed, therefore the total flow rate at the inlet is equal to the total flow rate at the outlet.

$$Q_{total} = \sum_i Q_{R_i} \tag{5.9}$$

The inventory levels of the product of interest is as follows.

$$\frac{dW_{P_j}}{dt} = \begin{cases} Q_{total} C_{P_j} - DR_{P_j}, \text{ if } Pur_{P_j} \geq 0.90\\ -DR_{P_j}, \text{ if } Pur_{P_j} < 0.90\end{cases} \tag{5.10}$$

where W_{P_j} is the inventory level, DR_{P_j} is the demand rate, and Pur_{P_j} is the purity level in the CSTR as defined in Eq. 5.11.

$$Pur_{P_j} = \frac{C_{P_j}}{\sum_j C_{P_j}} \tag{5.11}$$

The molar fractions of the reactant flow rates are defined in Eq. 5.12. Note that the molar fractions are utilized as the manipulated variables in the mpMPC control scheme, as demonstrated in the following sections.

$$\begin{aligned} a_{R_i} &= \frac{Q_{R_i}}{Q_{total}} \\ \sum_i a_{R_i} &= 1 \end{aligned} \tag{5.12}$$

Model approximation. The highly nonlinear nature of the model necessitates partitioning of the input space to capture the system dynamics with higher accuracy. Rigorous simulations of the high fidelity model suggests the partitioning of each degree of freedom available to the controller (i.e. a_{R_2} and a_{R_3}) to at least two mutually exclusive subspaces, respectively. Hence, the discrete time state space model generated in MATLAB® System Identification Toolbox™has the following form.

$$\begin{aligned} x_{t_c+1} &= A_c x_{t_c} + B_c u_{t_c} + C_c d_{t_c} \\ y_{t_c} &= D_c x_{t_c} \end{aligned} \tag{5.13}$$

$$\begin{aligned} u_{t_c} &= \left[u_{1,t_c}, u_{2,t_c}, u_{3,t_c}, u_{4,t_c}\right]^T \\ u_{1,t_c} &= a_{R_2},\ a_{R_2} \in [0, 0.5) \\ u_{2,t_c} &= a_{R_2},\ a_{R_2} \in [0.5, 1] \\ u_{3,t_c} &= a_{R_3},\ a_{R_3} \in [0, 0.55) \\ u_{4,t_c} &= a_{R_3},\ a_{R_3} \in [0.55, 1] \end{aligned} \tag{5.14}$$

where x_{t_c} is the identified states, u_{t_c} is the molar fractions of the reactant flow rates partitioned in the input space as given in Eq. 5.14, d_{t_c} is the total volumetric flow rate (Q_{total}), and y_{t_c} is the product concentrations (C_{P_j}). The state space matrices are given in the Supporting Information. Note that a_{R_1} is excluded from the manipulated variables due to the linear independence of the molar fractions.

Design of the mpMPC. The formulation of the mpMPC is based on Eq. 3.1 with additional soft constraints included as presented in Eq. 5.15. The tuning of the corresponding parameters is based on heuristic MPC design methods, and the parameters are provided in Table 5.3.

Table 5.3 Tuning parameters and objective weights of the mpMPC of the CSTR for Example 1 [95]

mpMPC design parameters	Value
N_c	6
M_c	2
QR	$\begin{bmatrix} 10^2 & 0 & 0 \\ 0 & 10 & 0 \\ 0 & 0 & 10 \end{bmatrix}$
$R1$	50
$P1$	90
$y_{\min}$	$[0, 0, 0]^T$
$y_{\max}$	$[1, 1, 1]^T$
$u_{\min}$	$[0, 0, 0]^T$
$u_{\max}$	$[1, 1, 1]^T$
$d_{\min}$	0
$d_{\max}$	500

$$
\begin{aligned}
\min_{u_{t_c}, z_{t_c}, \varepsilon_{t_c}} \quad & J(\theta) = \sum_{t_c=1}^{N_c} \|y_{t_c} - y_{t_c}^{SP}\|_{QR}^2 + \sum_{t_c=0}^{M_c-1} \|\Delta u_{t_c}\|_{R1}^2 + \sum_{t_c=1}^{N_c} \|\varepsilon_{t_c}\| \\
s.t. \quad & x_{t_c+1} = A_c x_{t_c} + B_c u_{t_c} + C_c d_{t_c} \\
& \hat{y}_{t_c} = D x_{t_c} \\
& y_{t_c} = \hat{y}_{t_c} + e \\
& e = y_{t_c=0} - \hat{y}_{t_c} \\
& x_{\min} \leq x_{t_c} \leq x_{\max} \\
& y_{\min} \leq y_{t_c} \leq y_{\max} \\
& u_{\min} z_{t_c} \leq u_{t_c} \leq u_{\max} z_{t_c} \\
& u_{\min} Y_{t_c} \leq u_{t_c} \leq u_{\max} Y_{t_c} \\
& d_{\min} \leq d_{t_c} \leq d_{\max} \\
& \Delta u_{\min} \leq \Delta u_{t_c} \leq \Delta u_{\max} \\
& - y_{*,t_c} + Pur_{\min} \sum_i y_{i,t_c} \leq -\varepsilon_{t_c} + \mathcal{M} Y_{t_c} \\
& 0 \leq \varepsilon_{t_c} \leq 1, \; z_{t_c} \in \{0, 1\} \\
& \theta = [x_{t_c=0}^T, y_{t_c=0}^T, d_{t_c=0}^T, (y_{t_c}^{SP})^T, u_{t_c=-1}^T, Y_{t_c}^T]^T \\
& \forall t_c \in \{0, 1, \ldots, N_c\}
\end{aligned}
\tag{5.15}
$$

where the additional terms ε_{t_c} is the slack variables, $P1$ is the penalty matrix, $Pur_{\min}$ is the minimum purity level required to trigger accumulation in the storage tanks, y_{*,t_c} is the concentration of the product of interest at time t_c, and $\mathcal{M}$ is a big-M parameter. The binary switch parameter Y_{t_c} determined by an upper level decision maker dictates the product of interest, and binary switch variable z_{t_c} determines the optimal input subspace. The soft constraints are constructed on y_{*,t_c} via slack variables ε_{t_c} to minimize the transition time by penalizing any production below the threshold purity level throughout the output horizon. Therefore, the non-negative slack variables ε_{t_c} contribute to the objective function if and only if the purity of the product of interest is below the threshold. Note that any process disturbances, such as reactant concentrations at the feed stream, can be easily incorporated in the control scheme without modifying the overall framework by simply introducing them as additional parameters.

The optimization problem given in Eq. 5.15 is reformulated as a mpMIQP problem and solved via the POP toolbox to generate the map of optimal control actions as affine functions of the system parameters.[2] The explicit expression of the control action is designed to (i) track a set point determined by an upper level decision maker, (ii) adapt proactively to changing operating modes (i.e. shifting between different products), and (iii) minimize the transition time by penalizing impure production periods.

Note that the mpMPC formulation utilizes a single state space model with piecewise affine inputs that are selected via binary switch variables, z_{t_c}. Therefore, the control scheme single-handedly recognizes the dynamics of the transitions between the production periods. Although the stability of the system under such transitions is left outside the scope of this study, it can be further investigated following the approach proposed by Grieder et al. [522].

Closed-loop validation. The control scheme is validated by exhaustive testing against the high-fidelity dynamic model under various scheduling decisions. Figure 5.4 presents a 2 h closed-loop operation with two distinct operational modes. The process starts from zero product concentration and goes through a shift from the production of P_3 to production of P_2 at $t = 60\ min$. This shift is manually enforced by changing the concentration set points from $y^{SP} = [0, 0, 0.1]^T$ to $y^{SP} = [0, 0.1, 0]^T$.

The closed-loop simulation in Fig. 5.4 validates the mpMPC as it (i) tracks the set points of three product concentrations, (ii) handles operations at different production modes, (iii) prioritizes purity satisfaction to minimize transition time, and (iv) maintains feasible operation by keeping the system within specified bounds. Note that the entirety of the closed loop simulations uses only one mpMPC scheme in both the production and the transition periods. Therefore, the controller parameters are tuned regarding every possible transition between the products.

High-fidelity model with the mpMPC embedded. The initial high-fidelity model given in Eqs. 5.7–5.12 is integrated with the derived control scheme in the form of Eq. 5.2.

[2] The complete solution of the problem along with the approximate model can be downloaded from http://paroc.tamu.edu.

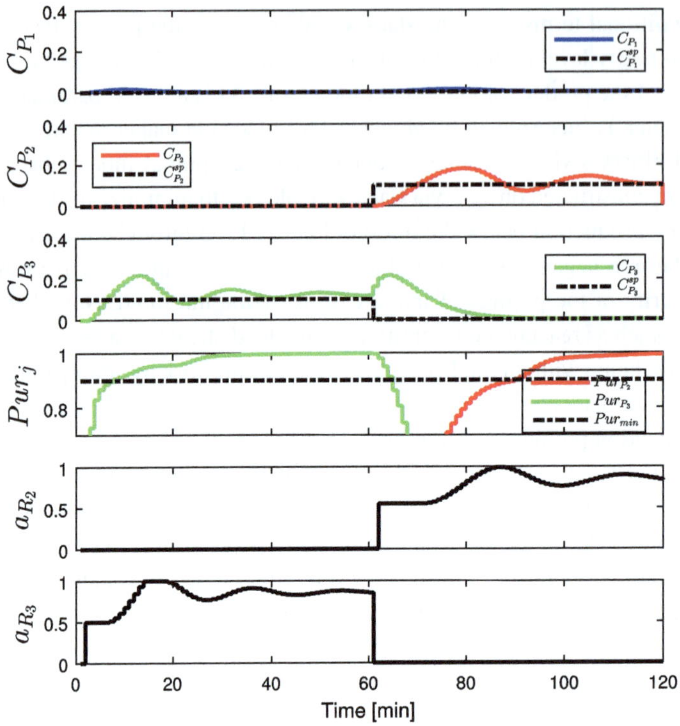

Fig. 5.4 Closed loop validation of the mpMPC against the high-fidelity model for Example 1 [95]

Model approximation. To keep the example tangible, only the mole balance around the storage tanks is considered in the upper level schedule, while the dynamics of the CSTR is accounted for in the lower level surrogate model formulation. The bilinear $Q_{total} C_{P_j}$ term in Eq. 5.10 results into a non-convex mpMINLP problem, for which only approximate solution algorithms exist. Hence, we postulate a mpMILP problem, for which POP toolbox features an exact algorithm, via replacing Eq. 5.10 with Eq. 5.16.

$$\frac{dW_{P_j}}{dt} = F_{P_j} - DR_{P_j} \tag{5.16}$$

where F_{P_j} is the molar product flow rate at the exit of the CSTR. Having merely linear terms in Eq. 5.16 enables the formulation of a mpMILP in the subsequent step.

The lower level surrogate model, on the other hand, is identified via MATLAB® System Identification Toolbox™ as described in the previous section. Three surrogate models are derived for three distinct operational modes (provided in the Supporting Information with their respective step and impulse responses).

Table 5.4 Parameters of the scheduler for Example 1 [95]

System parameters	Value
N_s	3
α [$/h.mol]	$[1.0, 1.5, 1.8]^T$
Δt_s [min]	60
$F_{\min}$	$[0, 0, 0]^T$
$F_{\max}$	$[50, 50, 50]^T$
$W_{\min}$	$[0, 0, 0]^T$
$W_{\max}$	$[50, 50, 50]^T$
$D_{\min}$	$[0, 0, 0]^T$
$D_{\max}$	$[60, 60, 60]^T$

Design of the scheduler and the surrogate model. The scheduler for this problem is designed to minimize the inventory cost, while satisfying continuous demand rates for the three products forecasted through the scheduling horizon.

$$
\begin{aligned}
\min_{F_{j,ts}, Y_{P_j,ts}} \quad & J(\theta) = \sum_{j=1} \sum_{t_s=1}^{N_s} \alpha_{P_j}^T W_{t_s} \\
s.t. \quad & W_{P_j,t_s+1} = W_{P_j,t_s} + \Delta t_s F_{j,t_s} - \Delta t_s DR_{P_j,t_s} \\
& \sum_{j=1} F_{j,t_s} = F_{total,t_s} \\
& \sum_{j=1} Y_{P_j,t_s} = 1 \\
& F_{\min} Y_{P_j,t_s} \leq F_{j,t_s} \leq F_{\max} Y_{P_j,t_s} \\
& W_{\min} \leq W_{P_j,t_s} \leq W_{\max} \\
& DR_{\min} \leq DR_{P_j,t_s} \leq DR_{\max} \\
& \theta = [W_{P_j,t_s=0}, DR_{P_j,t_s}]^T \\
& Y_{P_j,t_s} \in \{0, 1\}, \ \forall t_s \in \{0, 1, \ldots, N_s\}
\end{aligned}
\tag{5.17}
$$

where Y_{P_j,t_s} denotes the selected product P_j to be manufactured at time t_s, F_{j,t_s} is the molar product flow rate, Δt_s is the sampling time for the schedule. Note that Eq. 5.16 is discretized into time steps Δt_s.

The system parameters for Eq. 5.17 are given in Table 5.4.

The bridge between the mpMPC and the scheduler derived in Eqs. 5.15 and 5.17 is constructed based on Eq. 5.5. Analogous to the mpMPC, the surrogate model also features the soft constraints to enforce a threshold purity level.

$$
\begin{aligned}
\min_{Q_{total,tsm}, C^{SP}_{Pj,tsm}, \varepsilon'_{tsm}} \quad & J(\theta) = \sum_{t_{sm}=0}^{M_{sm}} + \| Q_{total,t_{sm}} - \tilde{Q}_{total,t_{sm}} \|^2_{R'} + \sum_{t_{sm}=1}^{N_{sm}} \| \varepsilon'_{t_{sm}} \| \\
s.t. \quad & Equations\ C.1{-}C.3 \\
& \tilde{Q}_{total,t_{sm}} = \frac{F_{total,t_{sm}}}{C_{P_*,t_{sm}=0}} \\
& y_{\min} \leq y_{t_{sm}} \leq y_{\max} \\
& Q_{\min} \leq Q_{total,t_{sm}} \leq Q_{\max} \\
& C^{SP}_{\min} \leq C^{SP}_{Pj,t_{sm}} \leq C^{SP}_{\max} \\
& - y_{*,t_{sm}} + Pur_{\min} \sum_i y_{i,t_{sm}} \leq -\varepsilon'_{t_{sm}} \\
& 0 \leq \varepsilon'_{t_{sm}} \leq 1 \\
& \theta = \left[x^T_{t_{sm}}, \frac{F_{total,t_{sm}}}{C_{P_*,t_{sm}=0}} \right]^T \\
& \forall t_{sm} \in \{0, 1, \ldots, N_{sm}\}
\end{aligned}
\tag{5.18}
$$

Note that the formulation given in Eq. 5.18 is only valid for the product of interest. Hence, three separate formulations are constructed for each product. Tuning of the surrogate model parameters is based on heuristic decisions that yield a desirable performance in the closed loop validation, and the parameters are given in Table 5.5.[2]

Closed-loop validation of the overall scheme. Closing the loop of the CSTR is performed via testing the scheduling and control scheme against the high fidelity model. Figure 5.5 presents a 12 h operation with the scheduler, the surrogate model, and the controller operating in tandem with the dynamic model while no specific knowledge of the demand profile assumed[3]. The scheduler (i) maintains low inventory levels, and (ii) adapts to the changes in the demand profile, while satisfying the continuous demand rate. Due to the rolling horizon strategy, the schedule is updated at every discretization step T_s, with the current inventory level and the new demand profile N_s time steps into the future. Note that the resultant production sequence is different from a cyclic schedule reported in Flores-Tlacuahuac and Grossmann [90] and Zhuge and Ierapetritou [153], since the demand rates in this case are time variant.

Figure 5.6 presents a snapshot of the first 6 h of operation, focusing on the lower level surrogate model decisions. The volumetric feed flow rate from the schedule and the surrogate model are juxtaposed in Fig. 5.6a to emphasize the corrective actions of the latter. During the transition between production regimes, the surrogate model saturates Q_{total} at its upper

[2] The complete solution of the problem along with the approximate model can be downloaded from http://paroc.tamu.edu.

[3] The explicit expressions of the simultaneous decisions at $t = 60$ min and $t = 105$ min are demonstrated in the Supporting Information.

Table 5.5 Parameters of the surrogate model for Example 1 [95]

System parameters	Model 1	Model 2	Model 3
N_{sm}	10	10	10
M_{sm}	1	1	1
T_{sm} [min]	6	6	6
R'	10^3	$\begin{bmatrix} 10^{-4} & 0 \\ 0 & 10^{-1} \end{bmatrix}$	$\begin{bmatrix} 10^{-4} & 0 \\ 0 & 10^{-1} \end{bmatrix}$
$P1'$	10^4	10^6	10^8
$y_{\min}$ [mol/L]	$[0, 0, 0]^T$	$[0, 0, 0]^T$	$[0, 0, 0]^T$
$y_{\max}$ [mol/L]	$[1, 1, 1]^T$	$[1, 1, 1]^T$	$[1, 1, 1]^T$
$Q_{\min}$ [L/min]	0	0	0
$Q_{\max}$ [L/min]	500	500	500
$C^{SP}_{\min}$ [mol/L]	$[0, 0, 0]^T$	$[0, 0, 0]^T$	$[0, 0, 0]^T$
$C^{SP}_{\max}$ [mol/L]	$[1, 1, 1]^T$	$[1, 1, 1]^T$	$[1, 1, 1]^T$

bound to purge the previous product left in the reactor. The transitions can also be monitored from the product purities presented in Fig. 5.6c. The surrogate model and the mpMPC operate in tandem to drive the system above the threshold purity level. The transitions to product P_2 specifically show that the integrated schedule and control scheme prioritizes the purity satisfaction to minimize the transition time.

Note the following:

- The explicit expressions for the optimal scheduling decisions enable rescheduling with a small computational cost when disruptive events occur in the product demands.
- The transition time is not determined explicitly by the integrated scheduling and control scheme, but is minimized through soft constraints in the surrogate model and controller formulations. The non-negative slack variables ε_{t_c} and $\varepsilon'_{t_{sm}}$ in Eqs. 5.15 and 5.18 are nonzero only if the product concentration of interest is below the threshold level, and contribute to the objective function $J(\theta)$ proportional to $P1$ and $P1'$, respectively. A more accurate approach would be allocating every time step for all products with binary variables to determine whether the purity threshold is satisfied. However, employing such a large number of binary variables in a multi-parametric programming problem results in an exponential increase in the computational burden. Hence, we alleviate this problem via the soft constraint formulation.
- The heavy penalty terms for purity satisfaction in the surrogate models result in steep changes in Q_{total} during the transitions, as observed in Fig. 5.6. The upper level schedule is unable to make such corrective decisions due to its large time step. The lower level

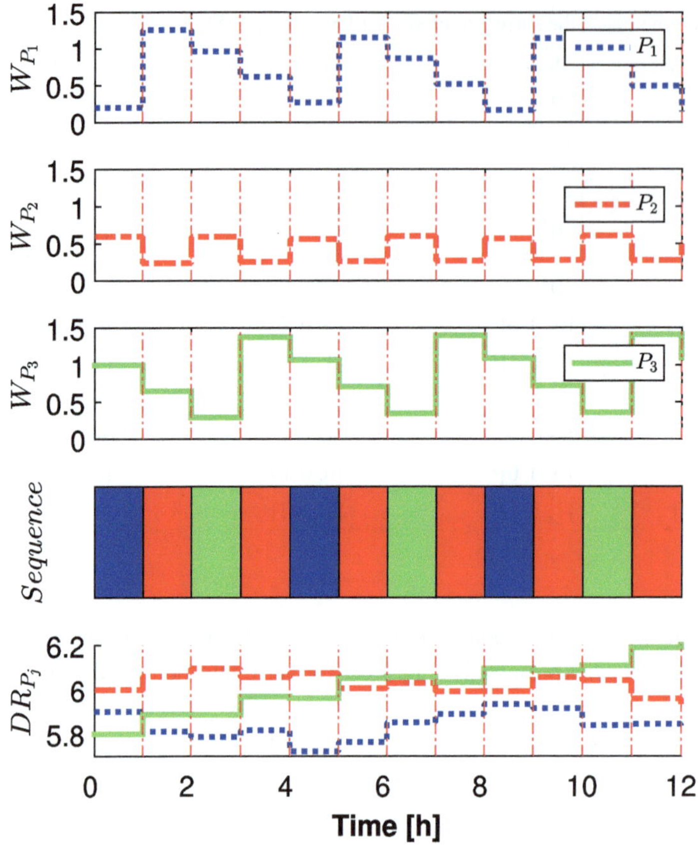

Fig. 5.5 Closed-loop validation of the scheduling scheme for Example 1 [95]

surrogate model provides time varying targets and set points for the controller due to the embedded closed loop dynamics in its formulation.

- Due to the strong nonlinearity of the high-fidelity model, the input space is partitioned as presented in Eq. 5.14. Finer partitions will yield more accurate controllers at the expense of increased computation time to generate the offline maps of optimal actions.
- Utilizing the maps of optimal solutions for the control, surrogate model, and schedule actions eases the online implementation. Calculation of the optimal actions is reduced from an online optimization problem to a simple look-up table algorithm and evaluation of an affine function.

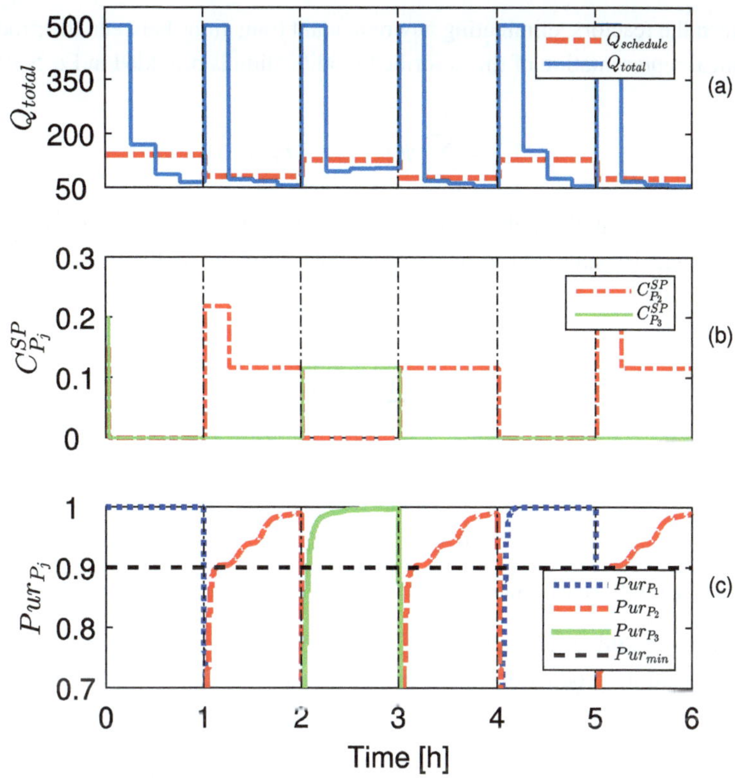

Fig. 5.6 Closed loop simulation of the CSTR for the first 6 h of operation for Example 1 [95]. **a** Volumetric flow rate determined by the scheduler, and the corrected action of the surrogate model, **b** Product concentration set points, **c** Product purities

5.4.2 Example 2 Two CSTRs Operating in Parallel

Problem definition. This case study extends the CSTR example from Sect. 5.4.1 to encompass two identical CSTRs operating in parallel. Due to the identical design of the reactors, the mpMPC and the surrogate model driving the closed-loop system are identical as well. Hence, the derivation of their formulations and attaining the explicit maps of solutions are omitted.

Cooperative operation of independent reactors requires a centralized scheduling scheme to allocate the production tasks on different reactors. However, the identical nature of the two closed-loop system dynamics creates a multiplicity of solutions, as the reactors are indistinguishable to the upper level schedule. Hence, the scheduling formulation presented in Eq. 5.17 is modified to (i) account for the previous production regime as an additional uncertain parameter, and (ii) penalize transitions between consecutive production regimes to break multiple solutions. Inclusion of the retrospective information provides a distinction between the reactors, eliminating any redundant transitions between the products. The mathematical representation of the described modification is provided in Eq. 5.19.

$$\Gamma_s = \sum_{t_s=0}^{N_s} \psi \left| Y_{P_j,t_s} - Y_{P_j,t_s-1} \right| \tag{5.19}$$

where ψ is a very small number that virtually penalizes the changes in the operational mode. Equation 5.19 can be reformulated as follows to maintain the linear structure of the scheduling problem.

$$\begin{aligned} \Gamma_s &= \sum_{t_s=0}^{N_s} \psi^T \bar{Y}_{t_s} \\ s.t. \quad & Y_{P_j,t_s} - Y_{P_j,t_s-1} \leq \bar{Y}_{t_s} \\ & - Y_{P_j,t_s} + Y_{P_j,t_s-1} \leq \bar{Y}_{t_s} \\ & 0 \leq \bar{Y}_{t_s} \leq 1 \end{aligned} \tag{5.20}$$

where $\bar{Y}_{t_s}$ is an auxiliary variable.

Design of the scheduler. The scheduling formulation presented in Eq. 5.17 is extended to account for multiple production lines, and modified with Eq. 5.20 to eliminate multiple solutions.

$$\begin{aligned} \min_{F_{j,t_s,l}, Y_{P_j,t_s,l}, \bar{Y}_{t_s,l}} \quad & J(\theta) = \sum_{j=1}^{N_s} \sum_{t_s=1} \alpha_{P_j}^T W_{P_j,t_s} + \sum_{l=1}^{N_{CSTR}} \sum_{t_s=0}^{N_s} \psi^T \bar{Y}_{t_s,l} \\ s.t. \quad & W_{P_j,t_s+1} = W_{P_j,t_s} + \sum_{l=1}^{N_{CSTR}} \Delta t_s F_{j,t_s,l} - \Delta t_s DR_{P_j,t_s} \\ & \sum_{j=1} F_{j,t_s,l} = F_{total,t_s,l} \\ & \sum_{j=1} Y_{P_j,t_s,l} = 1 \\ & Y_{P_j,t_s,l} - Y_{P_j,t_s-1,l} \leq \bar{Y}_{t_s,l} \\ & - Y_{P_j,t_s,l} + Y_{P_j,t_s-1,l} \leq \bar{Y}_{t_s,l} \\ & 0 \leq \bar{Y}_{t_s,l} \leq 1 \\ & F_{\min} Y_{P_j,t_s,l} \leq F_{j,t_s,l} \leq F_{\max} Y_{P_j,t_s,l} \\ & W_{\min} \leq W_{P_j,t_s} \leq W_{\max} \\ & DR_{\min} \leq DR_{P_j,t_s} \leq DR_{\max} \\ & \theta = [W_{P_j,t_s=0}, DR_{P_j,t_s}, Y_{P_j,t_s=-1,l}]^T \\ & Y_{P_j,t_s,l} \in \{0,1\}, \ \forall t_s \in \{0,1,\ldots,N_s\}, \forall l \in \{1,2,\ldots,N_{CSTR}\} \end{aligned} \tag{5.21}$$

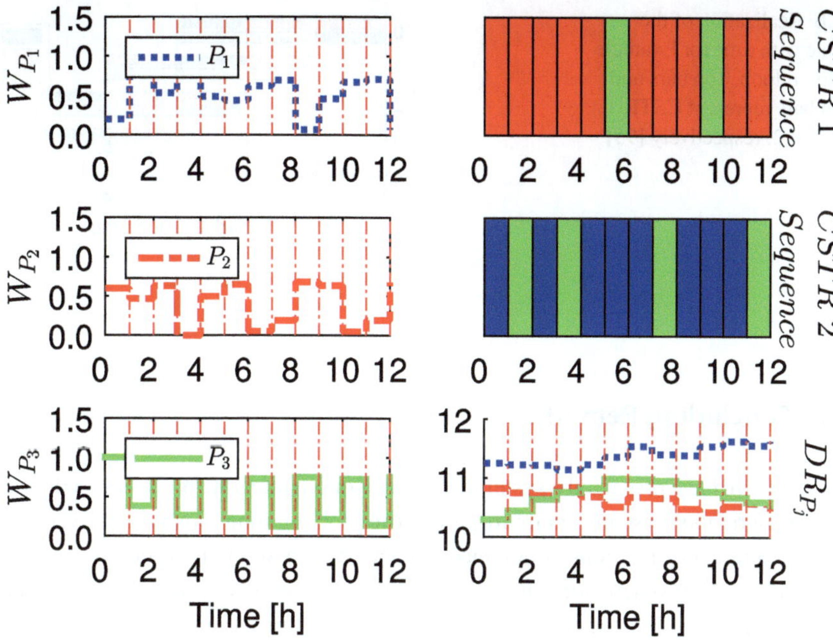

Fig. 5.7 Closed loop validation of the integrated scheduling and control scheme on two CSTRs for Example 2 [95]

where the additional weight ψ is tuned to be 0.001, and the number of the CSTRs, N_{CSTR}, is 2 by the problem definition.

Closed-loop validation. The validation of the overall scheduling and control scheme is presented in Fig. 5.7. The scheduler, the surrogate model, and the controller are operated in tandem with the high fidelity model for 12 h under a randomized demand profile. The integrated scheduling and control scheme delivers the additional task to coordinate multiple reactors to operate in parallel while satisfying the continuous demand rate. The inclusion of Eq. 5.20 in the objective function breaks the symmetry between the reactors and coordinates the production sequence. Consequently, uninterrupted manufacturing of the product of interest is maintained without redundant shifts between the reactors.

Figure 5.8 presents the evolution of the schedule with time for the first 4 h of operation. Note that the demand scenario is updated every hour in a rolling horizon manner, allowing rescheduling of the production sequence and the target quantities by utilizing the offline maps of optimal scheduling actions.

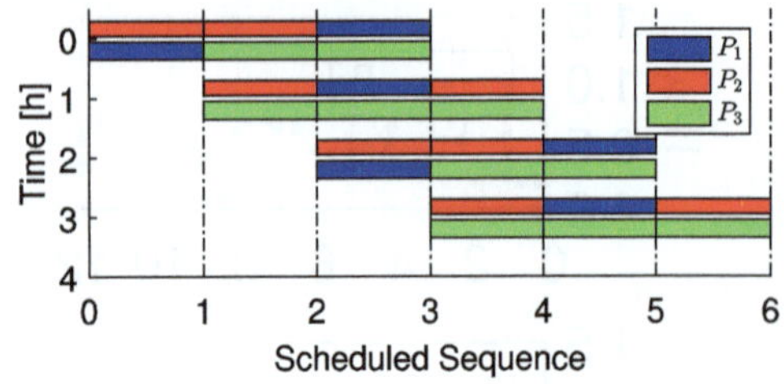

Fig. 5.8 Realization of the schedule with time for Example 2 in closed loop. Top bars and bottom bars represent CSTR 1 and CSTR 2, respectively [95]

5.5 Concluding Remarks

In this study, we have presented a systematic framework to integrate process scheduling and control in continuous systems via multi-parametric programming. We have derived optimal scheduling and control actions simultaneously based on a single high fidelity model. We take advantage of the synergistic interactions between the two decision making mechanisms to yield offline maps of optimal operations as explicit affine expressions at both long and short terms of a process. The generic structure of the framework renders it suitable for a software prototype towards enterprise-wide optimization.

This work aims to increase the operability, flexibility, and profitability of process systems through improving the scheduling and control decisions. Nevertheless, the processes with comparable capital and operating costs necessitates the consideration of the design aspect simultaneously with the scheduling and control. Hence, our current effort focuses on the unification of these three multi-scale problems.

Simultaneous Process Design, Scheduling, and Advanced Model Based Control

6

In this final chapter, we bring the components introduced in the earlier chapters together to integrate the process design, scheduling, and advanced control problems in a single unified framework. The framework features (i) a mixed-integer dynamic optimization (MIDO) formulation with design, scheduling, and control considerations, and (ii) a multi-parametric optimization strategy for the derivation of offline/explicit maps of optimal receding horizon policies. Explicit model predictive control schemes are developed as a function of design and scheduling decisions, and similarly design dependent scheduling policies are derived accounting for the closed-loop dynamics. Inherent multi-scale gap issues are addressed by an offline design dependent surrogate model. The proposed framework is illustrated by three example problems, a system of two continuous stirred tank reactor, which has been explored in the earlier chapters, a small residential combined heat and power (CHP) network, and a multipurpose batch process [7]. The scheduling problems in batch processes are typically much more complex compared to those in continuous processes due to increased number of tasks and a multitude of processing equipments. Therefore, we demonstrate the use of State Equipment Networks (SEN) in our framework. We also introduce a modeling technique to exponentially reduce the number of binary variables used in embedding the critical regions of the mpMPC strategy in the MIDO problem.

6.1 Overview

The complexity of decision making problems in the process industry has conventionally resulted in isolation of decisions with respect to the time scales of their effects on the operation, ranging from years-spanning supply chain management to seconds-long process control decisions. The isolated layers are structured hierarchically, as shown in Fig. 6.1a, with an information flow allowed dominantly in descending order in the time scales they span. However, independent and sequential assessment of the decision layers leads to subop-

B. Burnak et al., *Integrated Process Design and Operational Optimization via Multi-parametric Programming*, Synthesis Lectures on Engineering, Science, and Technology, https://doi.org/10.1007/978-3-032-10181-5_6

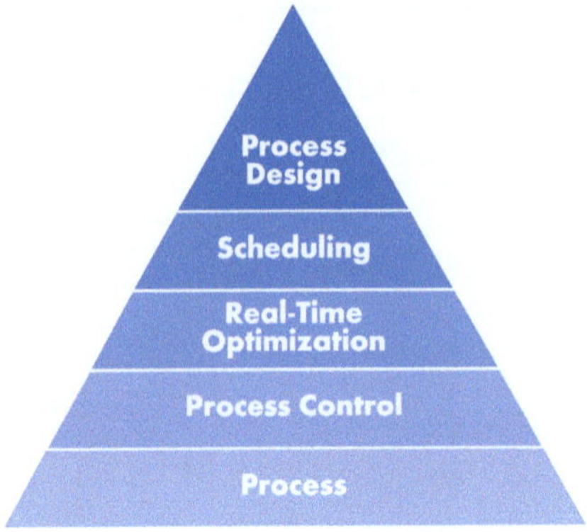

(a) Hierarchical approach.

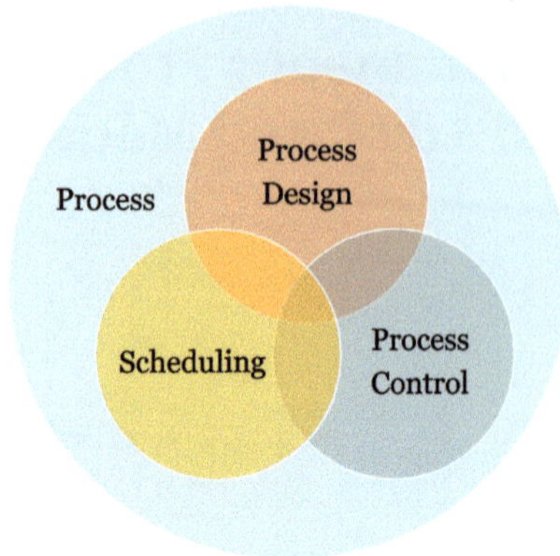

(b) Integrated approach.

Fig. 6.1 Decision making layers in an enterprise. The hierarchical approach is the conventional method that has been applied in the process industry. The aim of the integrated approach is to find decisions that is feasible and optimal in closed loop implementation [7]

timal, even infeasible operations. Integration of these layers across an enterprise is expected to deliver more profitable and reliable operations by benefiting from the synergistic interactions between different decisions [496]. Recent advances in operational research and rapid decrease in the cost of computational hardware provide an opportunity for the academia and the industry to seek a tractable and systematic methodology for simultaneous consideration of multi-scale decisions [147], as conceptualized in Fig. 6.1b. However, seamless integration of decision layers at different time scales and objectives is still an open question due to the high dimensionality and complexity of each constituent problem, such that process systems engineering tools and perspectives can play a key role for a holistic solution [523]. An indicative list of academic contributions towards the grand unification of process design, scheduling, and control is given in Table 6.1.

In this chapter, we present the complete framework to integrate design, scheduling, and control problems by deriving explicit maps of optimal decision making strategies at both levels of operation based on a single high fidelity model. We explicitly map the upper level layer decisions on the lower levels by multi-parametric programming. The explicit expressions at the lower level layers enable their representation in the upper level problems. In other words, the control problem is derived as a function of design and scheduling decisions, and similarly the scheduling decisions are design dependent, and aware of the controller dynamics. These explicit scheduling and control maps allow for an exact implementation in a design optimization problem. Furthermore, we introduce a design dependent surrogate model formulation to bridge the time scale gap between the schedule and the control problems, which is also solved offline. Direct inclusion of operating strategies in the design

Table 6.1 An indicative list of process design, scheduling, and control in literature [7]

Author (year)	Contribution
Design and control	
Perkins and co-workers (1991) [59], Kravaris and co-workers (1993) [414, 415], Pistikopoulos and co-workers (1994, 1997) [36, 395], Floudas and co-workers (1994, 2000, 2001) [397–399], Romagnoli and co-workers (1997) [400], Skogestad and co-workers (2014) [407], Gani and co-workers (1995) [411], Francisco and co-workers (2014) [413]	Feasibility, flexibility, stability, controllibility, resilience metrics in steady-state design optimization with MIDO or MINLP
Pistikopoulos and co-workers (2000, 2002, 2003a, 2003b, 2004) [71, 84, 228, 416–418, 421], Linninger and co-workers (2007) [80], Swartz and co-workers (2014) [419], Ricardez-Sandoval (2012) [386]	Integrated MIDO formulation/ decomposition with PID control or (mp)MPC
Biegler and co-workers (2007, 2008) [387, 424], Seider and co-workers (1992) [82], Ricardez-Sandoval and co-workers (2008, 2016, 2017) [72, 77, 425], Pistikopoulos and co-workers (1996) [35], Perkins and co-workers (2002, 2016) [75, 426], Flores-Tlacuahuac and co-workers (2009) [428], Barton and co-workers (2011, 2015) [430–432], Linninger and co-workers (2006) [434]	Iterative MINLP formulation with stochastic back-off formulation for uncertainty
Francisco and co-workers (2014) [470], Ricardez-Sandoval and co-workers (2009, 2011) [471, 472], Gani and co-workers (2012) [473], Mitsos and co-workers (2014) [474]	Review articles on design and control integration
Scheduling and control	
Grossmann and co-workers (2006, 2007, 2010, 2011, 2012, 2014) [90, 132, 184–186, 500], Gudi and co-workers (2010) [501], Biegler and co-workers (2012, 2014) [91, 109], Espuña and co-workers (2013) [506], You and co-workers (2013) [92], Baldea and co-workers (2016, 2018) [524, 525]	Decomposition of MIDO or MINLP and open loop optimal control
Allcock and co-workers (2002) [182], Pistikopoulos and co-workers (2003) [183], Ierapetritou and co-workers (2012) [153], You and co-workers (2012) [505], Baldea and co-workers (2015, 2018) [162, 526]	Formulation/Decomposition of MIDO schedule with PID control
Ierapetritou and co-workers (2014, 2018) [108, 115], Christofides and co-workers (2014a, 2014b, 2015, 2017) [193, 527–529], Baldea and co-workers (2015, 2018) [99, 115], Swartz and co-workers (2017) [194], Pistikopoulos and co-workers (2017, 2018) [95, 175], Hedengren and co-workers (2018) [530], Dua and co-workers (2019) [98]	(mp)MPC implementation in economic receding horizon policies
Puigjaner and co-workers (1995) [172], Marquardt and co-workers (2011) [192], Rawlings and co-workers (2011, 2012, 2013) [125, 128, 191], Pistikopoulos and co-workers (2013, 2014) [152, 509], Baldea and co-workers (2014) [127], Liu and co-workers (2016) [531]	Control theory/Economic MPC in scheduling problems
Reklaitis and co-workers (1996) [518], Grossmann (2005) [496], Harjunkoski and co-workers (2009) [498], Engell and Harjunkoski (2012) [88], Baldea and co-workers (2014) [181], Christofides and co-workers (2014) [104], You and co-workers (2015) [516], Ierapetritou and co-workers (2016, 2017) [166, 532]	Review articles on scheduling and control integration
Design, scheduling and control	
Grossmann and co-workers (2008) [3], Ricardez-Sandoval and co-workers (2015) [4]	Formulation of MIDO and open loop control under uncertainty
Ricardez-Sandoval and co-workers (2018) [6]	PI control and stochastic back-off approach for uncertainty
Pistikopoulos and co-workers (2018) [533]	Explicit, design dependent optimal rolling horizon strategies

optimization ensures the process operability by enforcing the decisions to be selected from the intersection of all layers from Fig. 6.1b.

The remainder of the chapter is organized as follows. Section 6.2 defines the integration problem that is addressed in this chapter. Chapter 6.3 brings the pieces of the PAROC framework introduced in the earlier chapters together for the complete integration of process design scheduling, and control. The framework is showcased in Sect. 6.4 on systems of reactors that were introduced in Chap. 5 and residential combined heat and power (CHP) units studied in Chaps. 3 and 4. Lastly, Sect. 6.5 presents concluding remarks and future directions.

6.2 Problem Definition

We consider a generic process where the interactions between the long term (design), middle term (schedule), and short term (control) decisions are sufficiently significant to impact the feasibility and the optimality of each individual decision. Therefore, we define the following problem that encapsulates all three decisions simultaneously.

(i) *Given*: A high fidelity model based on first principles or data-driven modeling techniques that accurately captures the dynamics of the system, any physical limitations of the system due to process safety considerations or product specifications, unit costs for design, raw material, energy, and inventory, revenue for unit product, and an accurate demand forecast.
(ii) *Determine*: Production sequence throughout an operating horizon, closed loop control strategy that delivers the product specifications, set points for the operation tailored for the dynamics of the closed loop strategy, size of the processing equipment that ensures operability of the process.
(iii) *Objective*: Minimize the operating and capital costs.

Note that the objective of the problem can be replaced by the minimization of the energy utilization, CO_2 emissions, processing time, or a combination of these tasks based on the application without changing the framework. In this study, we showcase the minimization of costs as it is the most frequently used objective in process operations.

6.3 Problem Formulation

A generalized mathematical representation of the simultaneous design, scheduling, and control problem can be formulated as a mixed integer dynamic optimization (MIDO) problem, as discussed in detail in Chaps. 1 and 2. The significantly different time scales of the manipulated variables yield a large scale MIDO problem that is computationally intractable by

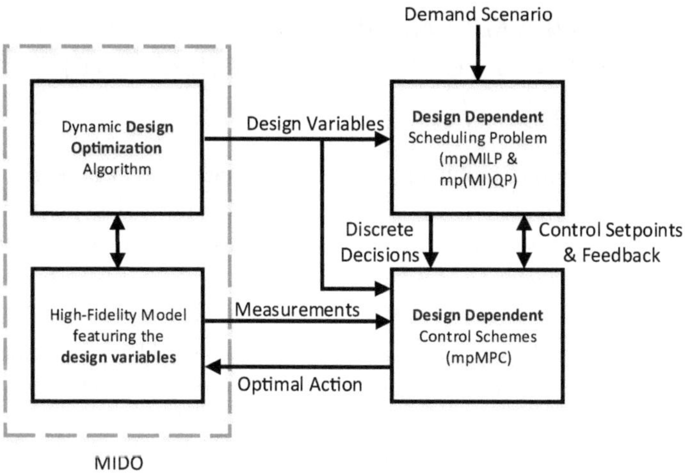

Fig. 6.2 Interactions between the process design, scheduling, and control layers [7]

the established approaches such as the direct, indirect, and dynamic programming based approaches. The proposed methodology comprises (i) developing an offline control policy that takes into account the different process dynamics stemming from the selection of the unit design and online economical decisions by following the procedure described in Chap. 4, (ii) deriving a scheduling policy based on the closed loop behavior of the system by following the procedure described in Chap. 5, and (iii) determining the design that minimizes the capital and operating costs for a given time period by utilizing the offline control and scheduling policies simultaneously. The interplay between the offline decision layers and the information flow in the overall MIDO formulation is illustrated in Fig. 6.2. The derivation of the design dependent explicit MPC and the offline scheduler has been previously summarized in Figs. 4.1 and 5.1, respectively.

The key difference of this chapter is the inclusion of both the scheduling and control decisions in the mpMPC formulation as parameters. Accounting for these decisions is achieved through developing the approximate models both as a function of design and scheduling variables, given by Eq. 6.1.

$$\begin{aligned} x^q_{t_c+1} &= A^q x^q_{t_c} + B^q u_{t_c} + C^q [d^T_{t_c}, s^T_{t_c}, des^T]^T \\ \hat{y}_{t_c} &= D^q x^q_{t_c} + E^q u_{t_c} + F^q [d^T_{t_c}, s^T_{t_c}, des^T]^T \end{aligned} \tag{6.1}$$

Note that Eq. 6.1 includes both the design and scheduling decisions as added parameters. Therefore, using this approximate model as the governing equations of the mpMPC, we derive the explicit optimal control law as given by Eq. 6.2.

$$u_{t_c}(\theta) = K_n\theta + r_n, \quad \forall\theta \in CR_n \tag{6.2a}$$

$$CR_n := \{\theta \in \Theta \mid L_n\theta \leq b_n\}, \quad \forall n \in \{1, 2, \ldots, NC\} \tag{6.2b}$$

$$\theta = [x_{t_c=0}^T, u_{t_c=-1}^T, d_{t_c=0}^T, s_{t_c}^T, des^T]^T \tag{6.2c}$$

Note that the scheduling problems will have the exact same parameters. The explicit optimal receding horizon policies are then incorporated in the MIDO problem, Eq. 2.1 for the complete simultaneous solution. Here, we would like to make some final remarks before demonstrating the framework on case studies.

Remark 1 Postulating all decision layers as optimization problems in the framework has practical benefits to be able to impose any physical limitations in each individual problem as hard or soft constraints. Such physical limitations can include safety considerations, thermodynamics, or operational policies. Implementation of these limitations is discussed and demonstrated in detail in the following examples.

Remark 2 The offline maps of optimal economical and operational decisions alleviate the computational burden of real-time optimization. During the online operation, we can simply determine the optimal actions *exactly* by a look-up table and affine function evaluations, instead of solving any optimization problems. On the other hand, determining the offline maps via multi-parametric programming and solving the integrated MIDO problem can be computationally expensive. However, these steps of the framework are evaluated once and completely offline.

Remark 3 The aim of the proposed framework is *not* to determine the global minimum of Eq. 2.1. Due to pre-postulation of scheduling and control strategies in the design optimization problem, Eq. 2.1 in fact describes a lower bound on the reconstructed MIDO. However, the reference trajectories acquired by Eq. 2.1 may be unattainable by the scheduling and control schemes when they are not explicitly accounted for, resulting into suboptimal, even infeasible operations. The proposed framework guarantees the operability of the system by properly embedding the operational strategies.

Remark 4 The proposed framework is *not* geared towards speeding up the computational time to solve Eq. 2.1. Because the solution profile and objective value can be suboptimal to Eq. 2.1 in the proposed framework (see Remark 3), the MIDO algorithm may terminate faster compared to the monolithic solution. In other words, any observed speed up in computational time is due to the search for a suboptimal but operable design, rather than an artifact of the solution strategy.

6.4 Case Studies

6.4.1 CSTR with Three Inputs and Three Outputs

In this case study, we extend the CSTR examples explored in Chap. 5 to also account for the design variables simultaneously. The reader is referred to Sect. 5.4 for the detailed problem description. The problem statement of the overall problem is given as follows:

(i) *Given*: A high-fidelity model of the three product CSTR, unit inventory costs, a functional expression for the CSTR fixed cost, a scenario of product demands.
(ii) *Determine*: Volume of the CSTR, production sequence, production rates, optimal reactant volumetric flow rates to achieve the target production rate and to reach the threshold purity.
(iii) *Objective*: Minimize the sum of operating and capital costs.

The objective in the problem definition can be achieved by determining the reactor design, production schedule, and closed loop dynamics that minimize the wasted raw materials and processing time. Therefore, (i) the controller is expected to deliver optimal transitions between all operating points determined by the scheduler, (ii) the scheduling decisions have to minimize the operating costs while accounting for the closed loop dynamics, and (iii) the reactor must be large enough to remain feasible throughout the entire operation, while avoiding overdesign to minimize the capital costs.

The complete MIDO formulation of the integrated problem is provided in (D.1). The following discussion breaks down the derivation and the solution strategy of the given multi-level MIDO problem.

Design-dependent, scheduling-aware mpMPC. We follow the procedure described in Chaps. 4 and 5. The developed mpMPC is validated against the high-fidelity model, under a range of scheduling decisions and design options. Figure 6.3 presents 4 h closed loop simulations for two reactor volumes ($V_1 = 0.4\,\text{m}^3$, $V_2 = 1.0\,\text{m}^3$). The process undergoes a step change from P_2 to P_3 after 2 h of operation to test the validity of the control scheme under different scheduling decisions and design configurations. Note that all operations are governed by a single explicit control law that is a function of the design and scheduling decisions.

The closed loop simulations presented in Fig. 6.3 shows that the developed control scheme is suitable for a range of scheduling and design options. The control scheme (i) achieves effective set point tracking for all three products simultaneously, (ii) minimizes transition time by prioritizing the purity satisfaction, (iii) recognizes the dynamics introduced by different scheduling decisions and design configurations, and (iv) maintains the operation within the inherent/imposed bounds of the system.

Design-dependent, control-aware scheduling. The developed design dependent and schedule-aware mpMPC is embedded into the original dynamic high fidelity model to

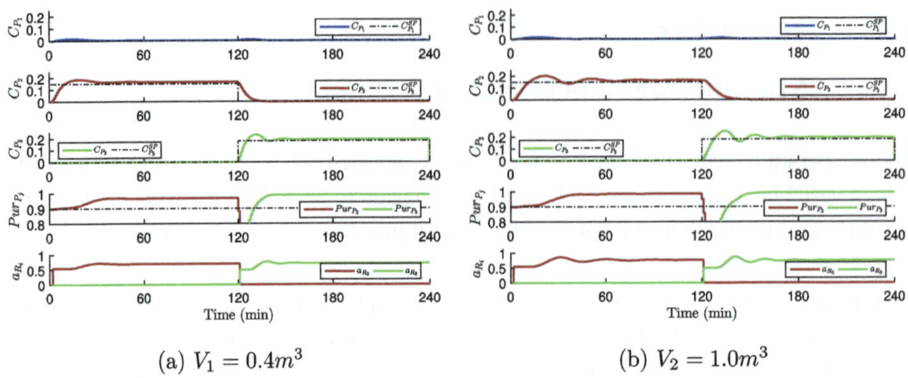

Fig. 6.3 Closed loop validation of the design dependent, schedule aware mpMPC with different design configurations for Example 1. The set points undergo a step change after two hours into the operation [7]

develop the scheduler in the form of an mpMILP and the surrogate model formulation as described in Sect. 5.4. After developing the explicit expressions for the scheduling problem, all three receding horizon policies are operated simultaneously on the high fidelity model under a range of design options and product demand variations. Figure 6.4 showcases the closed loop profiles for 12 h at the lower bound ($V_1 = 0.4\,\text{m}^3$) and the upper bound ($V_2 = 1.0\,\text{m}^3$) of the design range. Note that the same design dependent offline strategies are used in two reactors. The demand profiles for the products are randomly regenerated every hour, and the scheduling decisions are updated in a rolling horizon manner. The closed-loop simulations validate that the integrated scheduling and control scheme (i) maintains low inventory levels in the storage tanks, (ii) reactively adapts to changes in the demand profile, (iii) is applicable for a range of different design options. A sample of the offline scheduling and control decisions is demonstrated in Table 6.2, where a snapshot of the online operation of the large CSTR at $t = 5\,\text{h}$ is tabulated. Such explicit expressions are available for the range of design decisions, and will be used for design optimization described as follows.

Design optimization. The validated offline scheduling and control strategies are embedded in the overall MIDO problem given in Eq. 2.1 in the gPROMS environment. The capital investmentment for the reactor is determined by Eq. 6.3 [534].

$$C_e = a + bV^n \tag{6.3}$$

where C_e is the annualized reactor cost, and a, b, n are cost parameters given in Appendix D.3, along with the cost escalation indexes for year 2018. The minimum total annual cost is found as \$330 k/yr at $V = 0.69\,\text{m}^3$. Note that the scheduling and control strategies yield feasible operation for the optimal reactor volume as a result of their design dependence. Therefore, treating the design, scheduling, and control problems simultaneously ensures the

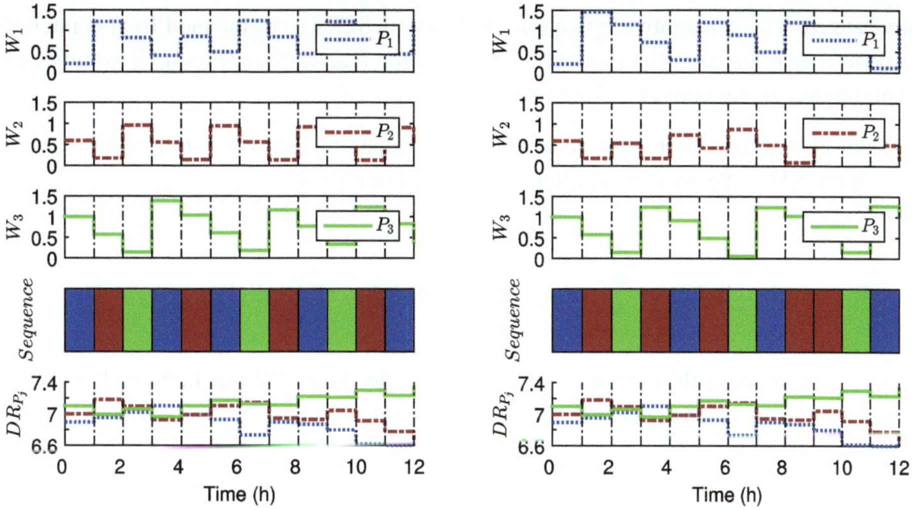

(a) $V_1 = 0.4m^3$ (b) $V_2 = 1.0m^3$

Fig. 6.4 Closed loop validation of the integrated scheduling and control scheme operating in tandem, demonstrated on two reactors with different volumes [7]

Table 6.2 A sample for the optimal design dependent scheduling and control decisions at $t = 5$ h for the large CSTR ($V_2 = 1.0\,\text{m}^3$) [7]

Decision variable	Affine expression
$F_{3,t=0}$	$= -16.7W_3 + DR_{3,t=0} + DR_{3,t=1h} + DR_{3,t=2h}$
$F_{2,t=1h}$	$= -16.7W_2 + DR_{2,t=0} + DR_{2,t=1h} + DR_{2,t=2h}$
$F_{1,t=2h}$	$= -16.7W_1 + DR_{1,t=0} + DR_{1,t=1h} + DR_{1,t=2h}$
$Q_{total,t=0}$	$= 500$
$CP^{SP}_{2,t=0}$	$= 0.91(CP_{1,t=0} - 0.003) - 0.007(CP_{2,t=0} - 0.14) - 0.12$
a_1	$= 0.45 - 6 \times 10^{-3}V$
a_2	$= 0.55 + 6 \times 10^{-3}V$

operability of the system, as the MIDO problem comprises the exact closed loop strategies that will be used online during the operation.

6.4.2 Two CSTRs Operating in Parallel

This case study presents an extension of the single CSTR example discussed in Sect. 6.4.1 to two CSTRs operating in parallel. The exact same control strategy and the surrogate model formulations are employed because the open loop dynamics of the system remains unchanged. The cooperative operation of the two CSTRs is maintained by a centralized scheduler that allocates the production tasks on the reactors based on their volumes and their production regimes at a given time.

Design of the scheduler. The governing approximate model is generalized to represent multiple CSTRs operating in parallel, as presented in Eq. 6.4.

$$Wj,t+1 = Wj,t + \sum_{p=1}^{N_{CSTR}} \Delta t F_{j,t,p} - \Delta t DR_{j,t} \quad \forall j, \forall t \in \{1,\dots,N_s - 1\} \tag{6.4}$$

where the number of the reactors, N_{CSTR}, equals 2 by the problem definition. The product assignment constraints are also generalized as presented in Eq. 6.5.

$$\begin{aligned} &\sum_{j=1} y_{j,t,p} = 1 \\ &\underline{F} y_{j,t,p} \le F_{j,t,p} \le \overline{F} y_{j,t,p} \end{aligned} \tag{6.5}$$

Closed loop validation. The generalized offline scheduling scheme is validated against the high fidelity model of the two reactor system. Figure 6.5 showcases a scenario with one small reactor ($V_1 = 0.4\,\mathrm{m}^3$) and one larger reactor ($V_2 = 1.0\,\mathrm{m}^3$) operated in parallel. The integrated scheduling and control scheme is able to drive the inventory level of the most costly product, W_{P_3}, close to zero by assigning it to the larger reactor. The large reactor is capable of satisfying the demand on P_3 standalone, and the small reactor has a faster transition rate because of the lower retention time. Therefore recognizing the closed loop dynamics and the capacity of the reactors, the integrated schedule assigns the costly product, P_3, to the large reactor, and alternates the production between P_1 and P_2 in the small reactor.

Design optimization. The offline maps of scheduling and control are embedded in the overall MIDO problem in the gPROMS environment. The reactor configuration with volumes $V_1 = 0.44\,\mathrm{m}^3$ and $V_2 = 0.92\,\mathrm{m}^3$ minimizes the total annual cost accounting for the capital and operating costs. Note that one large reactor and one small reactor is selected to deliver (i) uninterrupted production of one of the products depending on their unit storage prices and demand rates throughout the horizon, and (ii) fast transitions for alternating production of the remaining products, respectively.

6.4.3 Residential Combined Heat and Power (CHP) Unit

This case study extends the combined heat and power generation system (CHP) explored in Chaps. 3 and 4 to simultaneously account for the design decisions, which is the volume of the internal combustion engines in the CHP units. In this case study, we take into account the external factors that affect the desired level of operation, i.e. the fluctuations in the heat and

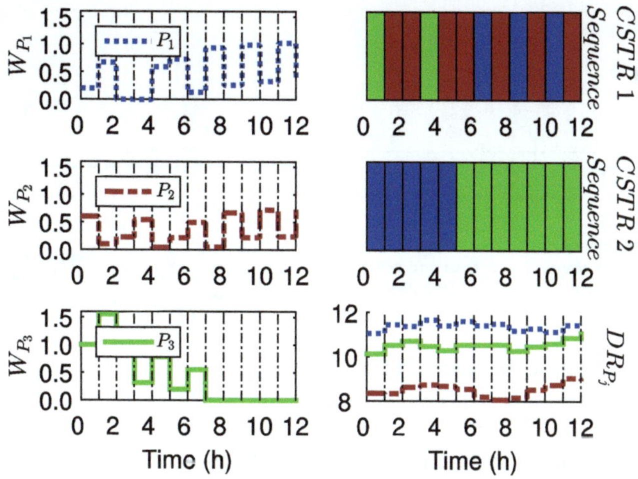

Fig. 6.5 Closed loop validation of the scheduling scheme generalized for two reactors operating in parallel [7]. The volumes of the reactors are $V_1 = 0.4\,\text{m}^3$ and $V_2 = 1.0\,\text{m}^3$, respectively

power demand rates, and changing market prices for the electricity and fuel. We consider a residential district with 10 units, all of which are supplied hot water for heating purposes and electricity from a single CHP unit. The hot water can be stored in a buffer tank if the produced heat content exceeds the demand rate. Additional electricity can be supplied from the central grid if the CHP unit falls short, and a supplementary boiler is assumed to be available at all times to provide more heat content. Excess electricity produced from the CHP unit can be sold to the central grid for revenue, and excess hot water can be disposed of at an expense. Note that the rapidly changing electricity prices in day time and night time has a significant economic impact on the operation of a CHP unit. For instance, it may be more profitable to operate the CHP unit at a higher capacity during the day time because of the increased cost of electricity purchase, and at a lower capacity at the night time when the cost decreases. Therefore, determining the most cost effective operation can be achieved by taking into account the fluctuation in the prices, demands rates, as well as the dynamics of the CHP units. A generalized flowsheet of the CHP system with two parallel CHP units is presented in Fig. 6.6. However in this section, we focus on a system with a single CHP system supplying the heat and power to the residential units. Parallel operation of multiple units will be discussed in the subsequent example.

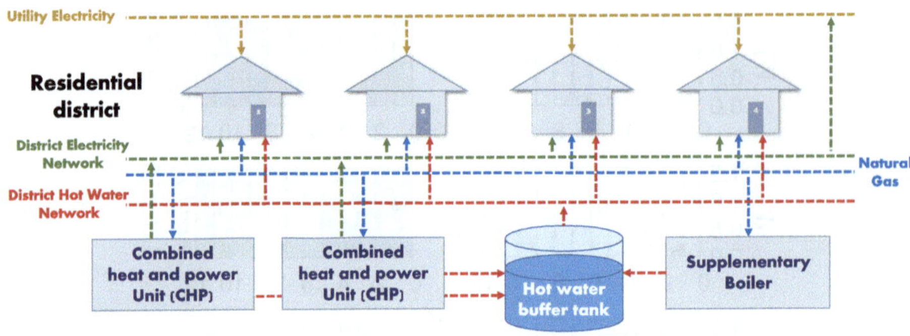

Fig. 6.6 Flowsheet of the CHP system and residential units [7]

The problem statement of the case is given as follows:

(i) *Given*: A high-fidelity model of the CHP, a demand scenario for electricity and heat consumption, investment cost of the CHP unit as a function of its size, market prices of fuel and purchasing/selling electricity.
(ii) *Determine*: Internal combustion engine (ICE) size of the CHP, a schedule for the transactions with the grid and fuel purchases, operating level of the CHP.
(iii) *Objective*: Minimize the sum of operating and capital costs.

The size of the ICE directly affects the process time of the system, and thus the responsiveness of the CHP to fluctuations in the demand rates and market prices. ICEs smaller in size have lower transition time, hence they can deliver fast responses to changes in the operating set points. On the other hand, larger ICEs can supply more power and heat to the residential units when the demand rates are high. The trade-off between the responsiveness and the capacity of the CHP is addressed by integrating a design dependent scheduler and controller in the design optimization problem.

High-fidelity dynamic model. There are two main components taken into account in the CHP model, (i) a natural gas powered ICE to produce electrical power, and (ii) a cooling system that recovers the excess heat content of the ICE. We also include the dynamics of the throttle valve that manipulates the inlet air mass flow rate, and the intake manifold that distributes the air into the ICE cylinders. For the detailed mathematical model, the reader is referred to Diangelakis et al. [377].

Model approximation. The original high fidelity model is a DAE system with 364 algebraic and 15 differential relations in the continuous domain. In our previous studies, the complexity of the overall system is addressed by decomposition into two approximate models, namely a power production subsystem and a heat recovery subsystem [85, 535]. The former operating mode gives the relation between the throttle valve opening and the power output of the CHP, while the latter is used to estimate the water temperature at the outlet as a function of the power output and the water flow rate into the heat recovery system. Equation 6.6 presents the identified state-space model for the power production subsystem.

$$\begin{aligned} x_{t+1} &= 0.9799x_t + 0.0006u_t + 6.516V \\ y_t &= 7.839x_t \end{aligned} \tag{6.6}$$

where x_t is the identified state, u_t is the throttle valve opening, V is the volume of the ICE, y_t is the electrical power generated by the CHP.

The heat recovery subsystem is an explicit function of the output of the power production subsystem and is given in Eq. 6.7.

$$\begin{aligned} x_{t+1} &= \begin{bmatrix} 0.997 & 0.103 & -0.003 \\ -0.002 & 0.940 & 0.116 \\ -0.058 & -0.056 & 0.179 \end{bmatrix} x_t + \begin{bmatrix} -0.008 & 0.001 \\ 0.280 & 0.033 \\ -1.280 & 0.146 \end{bmatrix} u_t \\ y_t &= \begin{bmatrix} -529.9 & -2.827 & 0.252 \end{bmatrix} x_t \end{aligned} \tag{6.7}$$

where x_t is the set of identified states, u_t are the the power generation level and water flow rate, respectively, and y_t is the temperature of the hot water at the outlet. The discretization time steps of the models presented in Eq. 6.6 and 6.7 are both 0.1 s.

Design of the mpMPC. The two subsystems derived in the previous step are operated by a decentralized control policy, which comprises interlinked control strategies for each subsystem. We define two operational modes for the decentralized control policy defined as follows.

- *Electricity driven mode (Mode 1)*: The operating level of the CHP, i.e. the power set point, is determined based on the power demand. Therefore, the throttle valve opening is manipulated primarily to satisfy the demand on electricity. The operating level projected by the electricity generation subsystem is treated as a measured disturbance by the heat recovery subsystem, hence the produced heat is a function of the power output of the CHP. The heat production level of a standalone CHP unit can be insufficient to satisfy the heat demand at a given time, requiring the use of the supplementary boiler. It is also possible that the produced heat content exceeds the heat demand, in which case the hot water is stored in a buffer tank.
- *Heat production driven mode (Mode 2)*: The operating level of the CHP is determined based on the heat demand. Tracking a water temperature set point at 70 °C, heat recovery subsystem (i) determines an operating level set point to ensure sufficient heat production by the power production subsystem, and (ii) manipulates the cooling water flow rate to recover enough heat to satisfy the demand. Analogous to mode 1, the power production level may not match the electricity demand. In case of insufficient power, additional electricity is purchased from the central grid, and excess electricity is sold back to the grid for revenue.

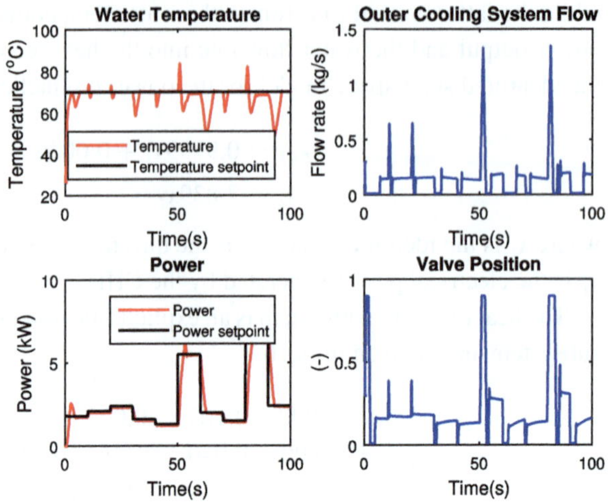

Fig. 6.7 Closed loop simulation of a CHP unit with $V = 1500$ cc, operated with mode 1 [7]

The reader is referred to Diangelakis et al. [85, 86, 535] for more details on the operating modes and a quantified evaluation of the decentralized control policy.

Note that changing the operating modes creates an offset between the new set point and the current output of the system. This offset has economical consequences on the operation and dictates the quantity of electricity purchases/sales, usage of the buffer tank and the supplementary boiler. These economical aspects are addressed and mitigated in the scheduling formulation.

Closed loop validation. The design dependent decentralized control policy is validated against the high fidelity model under a range of different design and scheduling decisions. Figure 6.7 shows a closed loop simulation of a CHP with $V = 1500$ cc operated with mode 1 only. The power set point is subject to random changes throughout the operation.

Similarly, closed loop simulation on a larger CHP ($V = 5000$ cc) is demonstrated in Fig. 6.8. Note that due to operating mode 2, the power set point is subject to changes dictated by the heat recovery subsystem.

High fidelity model with the mpMPC embedded. The explicit form of the decentralized control policy is implemented in the original high fidelity model.

Model approximation. The closed loop high fidelity model is used to develop an approximate model for the scheduler via the MATLAB System Identification Toolbox. The identified model establishes a relation between the power production and heat storage levels, and the change in the power production set point, as presented in Eq. 6.8.

$$\begin{bmatrix} E_{t+1} \\ B_{t+1} \end{bmatrix} = \begin{bmatrix} 0.999 & 0 \\ 37.9 & 0.955 \end{bmatrix} \begin{bmatrix} E_t \\ B_t \end{bmatrix} \begin{bmatrix} 99.5 & 0 & 0 \\ 0 & 11.2 & -11.2 \end{bmatrix} \begin{bmatrix} R_t \\ Q_t \\ D_t \end{bmatrix} + \begin{bmatrix} 0 \\ -11.2 \end{bmatrix} \zeta_t^h \tag{6.8}$$

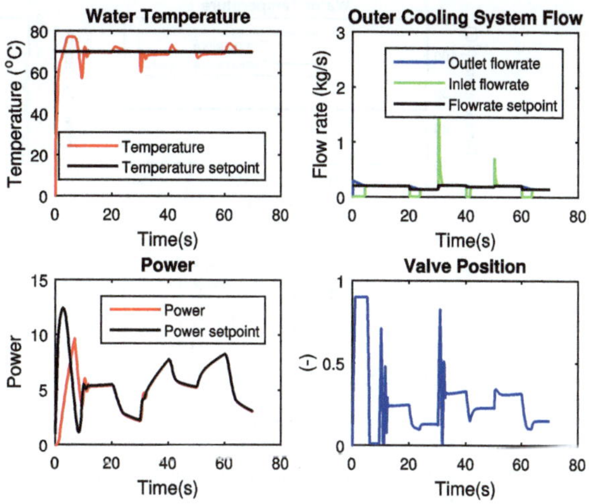

Fig. 6.8 Closed loop simulation of a CHP unit with $V = 5000$ cc, operated with mode 2 [7]

where E_t is the energy production level, B_t is the heat storage level, R_t is the change in the power production set point, Q_t is the additional heat supplied from the boiler, D_t is the disposed heat, ζ_t^h is the heat demand, and the time step of the model is 10 s. We also use an overall energy balance for the relation between the power production, power demand, and electricity purchases from the central grid, presented in Eq. 6.9.

$$P_t + E_t = \zeta_t^p + W_t \tag{6.9}$$

where P_t is the electricity purchase, ζ_t^p is the power demand, and W_t is the excess electricity sold back to the grid.

Design of the scheduler. The objective of the schedule is to minimize the operating costs, including energy production, energy purchases and sales, and inventory costs, as given in Eq. 6.10.

$$\sum_{t=1}^{N_s} \beta E_t + \psi_t P_t - \nu_t W_t + \xi_t Q_t + \omega_t D_t + \gamma B_t \tag{6.10}$$

where the Greek letters denote the corresponding cost parameters. Note that the CHP unit is assumed to be operational throughout the scheduling horizon. Hence, on/off switching costs are excluded in the objective function. This assumption will be relaxed in Sect. 6.4.4 where we discuss a parallel operation of multiple CHP units. The objective function is subject to the approximate CHP model derived in Eqs. 6.8 and 6.9, as well as the lower and upper bounds on the optimization variables.

The power production capacity of the CHP unit is a function of the ICE size (i.e. $\overline{E} = \overline{E}(V)$). The schedule treats this design variable as a bounded parameter along with the initial conditions of the system, power and heat demands, unit cost of purchasing fuel and power, and unit revenue of selling power, as listed in Eq. 6.11.

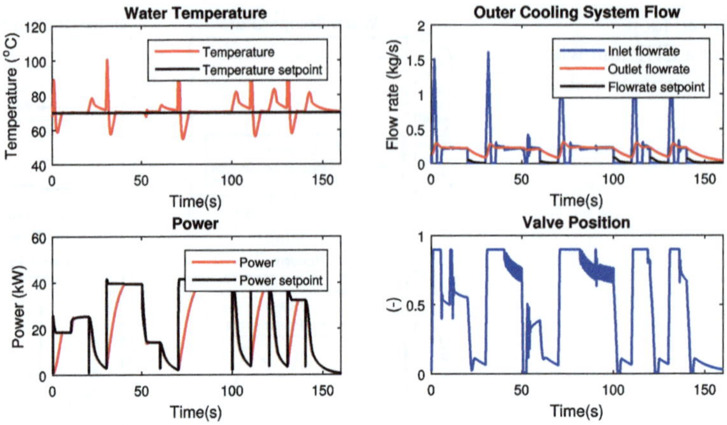

Fig. 6.9 Closed-loop validation of the integrated scheduling and control scheme on a CHP with $V = 5000\,\text{cc}$ [7]

$$\theta = [V, E_t, B_t, \zeta_t^h \zeta_t^p, \beta_t, \psi_t, \nu_t, \xi_t, \omega_t, \gamma_t] \tag{6.11}$$

Design of the surrogate model. Equations 6.8 and 6.9 are resampled in the time steps of the controller, and substituted in the surrogate model formulation presented in Eq. 5.5. The resampled state space matrices are given in Appendix D.2.

Closed-loop validation. The integrated scheduling and control scheme is validated against an extensive set of design options and demand profiles. Figure 6.9 shows a snapshot of a closed loop simulation of a CHP unit with a volume $V = 5000\,\text{cc}$. Note that the power set point throughout the operation is determined by the offline schedule, and translated into the time steps of the controller by the surrogate model.

Design optimization. We formulate a MIDO problem in the gPROMS environment using the high fidelity model, the explicit design dependent relations for the scheduler, surrogate model, and the controller. The capital investment cost is assumed to be a linear function of V, and is given in Appendix D.3. A CHP unit with an ICE volume of $V = 1710cc$ yields the scheduling and control strategies that minimizes the total annualized cost that includes the capital and operating costs.

6.4.4 Two CHPs Operating in Parallel

The single CHP case study presented in Sect. 6.4.3 is extended to include two CHP units operating in parallel. We generalize the scheduling formulation to account for multiple CHP units, and showcase the proposed algorithm on a system with two units. We also include the dynamics stemming from switching on/off the units, and their impact on the operational optimization.

Design of the scheduler. Evidently, multiple CHP units have a greater capacity to supply heat and power compared to a single unit. However, the total production rate of multiple units can exceed the demand rates significantly even when they are operated at their lowest capacities. In other words, operating one CHP unit standalone can be more cost effective than operating two CHPs simultaneously at low demand rates. Therefore, we include the start-up and shut-down dynamics in the schedule to account for the trade-off between switching on/off the operation and maintaining the operating status of a unit.

The cost of switching on/off is described by Eq. 6.12.

$$\sum_{i=1}^{N_{CHP}} \sum_{t=1}^{N_s} \phi_i S_{i,t} + \pi_i F_{i,t} \tag{6.12}$$

where N_{CHP} is the number of CHP units, $S_{i,t}$ and $F_{i,t}$ are binary variables that indicate the start-up and shut-down status, and ϕ_i and π_i are their unit costs, respectively. The impact of the switching status variables is incorporated in the schedule by introducing lifting-state variables, $\tilde{S}_{i,t,n}$ and $\tilde{F}_{i,t,n}$, as presented in Eq. 6.13.

$$\begin{aligned} \tilde{S}_{i,t+1,n} = \tilde{S}_{i,t,n-1}, \quad \tilde{S}_{i,t,n=0} = S_{i,t} \\ \tilde{F}_{i,t+1,n} = \tilde{F}_{i,t,n-1}, \quad \tilde{F}_{i,t,n=0} = F_{i,t} \end{aligned} \tag{6.13}$$

The state lifting-variables determine the operating status of the CHP units as described in Eq. 6.14.

$$\begin{aligned} S_{i,t} - F_{i,t} &= X_{i,t} - X_{i,t-1} \\ X_{i,t} &\geq \sum_{n=0}^{\delta_i^{up}} \tilde{S}_{i,t,n} \\ 1 - X_{i,t} &\geq \sum_{n=0}^{\delta_i^{dn}} \tilde{F}_{i,t,n} \end{aligned} \tag{6.14}$$

where $X_{i,t}$ is a binary variable that indicate the operating status, δ_i^{up} and δ_i^{dn} are the start-up and shut-down times of the ith CHP unit. The interested reader is referred to Subramanian et al. [128] for more details on scheduling with lifting-state variables, and to Kopanos et al. [152] for an application of reactive scheduling using lifting-state variables on a CHP system.

The cost function given in Eq. 6.10 is generalized to encapsulate the operating cost of multiple CHP units, as presented in Eq. 6.15.

$$\sum_{i=1}^{N_{CHP}} \sum_{t=1}^{N_s} \beta E_{i,t} + \sum_{t=1}^{N_s} \psi_t P_t - \nu_t W_t + \xi_t Q_t + \omega_t D_t + \gamma B_t \tag{6.15}$$

The objective function of the schedule comprises the operating and purchasing costs described by Eq. 6.15 and the switching costs given in Eq. 6.14.

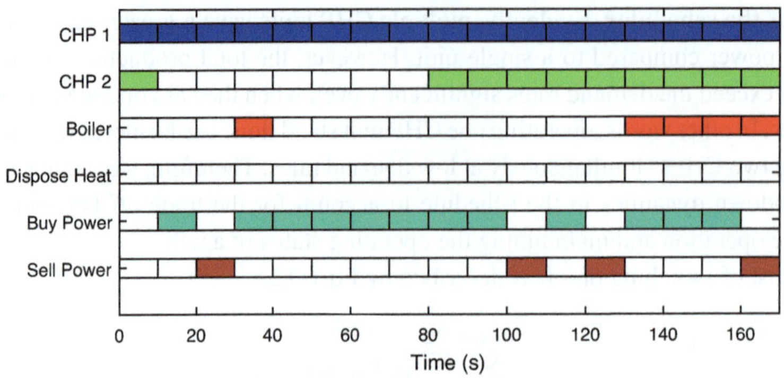

Fig. 6.10 Closed loop simulation of the generalized scheduling scheme in two CHP units operating in parallel. The volumes of the ICE are $V_1 = 1500$ cc and $V_2 = 4500$ cc, respectively [7]

Fig. 6.11 Snapshot of the electricity and heat demand profiles. Note the steep increase in demand in short notice [7]

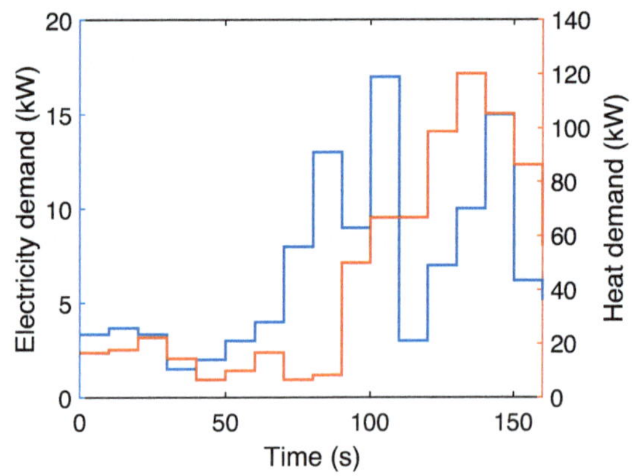

Closed loop validation. The developed scheduling strategy is implemented on the high fidelity model and operated in tandem with the offline controller. Figure 6.10 shows a snapshot of the scheduling level decisions of an operation with two CHP units with ICE volumes $V_1 = 1500$ cc and $V_2 = 4500$ cc, under a rapidly escalating demand profile given in Fig. 6.11. The following are some observations and remarks on the closed loop performance of the developed scheduling and control strategies.

- The small CHP is operated standalone at low demand rates.
- The large CHP is operated when either of the demand rates are high.
- Due to the time loss during the start-up of the large CHP, grid electricity is used to supplement the deficit.
- The recovered heat content is not wasted by disposal.

- In both CHP units, set point tracking is achieved via the same design-dependent control strategy, which is developed and discussed in Sect. 6.4.3. The closed loop profiles of water temperature, power output, cooling water flow rate, and valve position are omitted here fore brevity.

Design optimization. The MIDO problem is formulated by embedding the offline scheduling and control schemes in the high fidelity model in the gPROMS environment. A CHP system with ICE volumes of $V_1 = 2050$cc and $V_2 = 2700$ cc yields the most cost effective scheduling and control strategies, minimizing both the capital and operating costs. Note that one small CHP unit is selected to be operated continuously even at low demand rates, and one larger CHP unit to be operational under higher demand rates, a similar outcome of the case study presented in Sect. 6.4.2.

6.4.5 Multipurpose Batch Process

The goal of this example is to present the applicability of the unified theory and framework on batch processes, where the scheduling problem is considerably more complex and requires more sophisticated modeling representations. Here, we consider a multipurpose batch process where the products are allowed to follow different routes through the plant at different times [536]. The objective of these batch plants may vary depending on the application, such as minimizing the cost, minimizing the makespan, or maximizing the yield of a specific product. Therefore, the problem statement is outlined as follows.

Given. First principle dynamics to manufacture the desired products (preprocessing, reaction, separation), any physical limitations regarding the product quality and process safety, unit capital and operating costs, and the range of demands on products.

Determine.

1. Process design decisions: Dictates the capacity of the processing units.
2. Process scheduling decisions: Includes task allocation, production span or cycle, production sequence, and batch sizes.
3. Real-time optimization decisions: Input and output trajectories that are transmitted to the regulatory controller.
4. Closed-loop control decisions: Model predictive control strategy that governs the process through the control instruments.

Objective. Minimize cost, minimize, makespan, maximize yield, etc.

Unlike the previous examples, the process flow is not predetermined due to availability of multiple processing units that can deliver the same task with different efficiencies. There are multiple scheduling representations proposed in the literature that systematically accounts for the trade-offs between the scheduling decisions. State-Task Network (STN) [65], Resource-Task Network (RTN) [537], and State-Equipment Network (SEN) [91] are the most widely used scheduling techniques that provide a systematic modeling framework and solution strategy for these decisions through mixed-integer linear programming (MILP). In this example, we utilize the SEN framework for the scheduling problem due to its suitability for the integration with dynamic models and the optimal control problem [91]. Therefore, we first introduce the essential components of the SEN framework with the logical disjunctions to incorporate in the integrated dynamic optimization problem. We also discuss a base-2 numeral system based approach to embed the critical regions of the controller. This system does not provide a theoretical improvement to the optimality of the solution, but leverages the computational complexity by exponentially decreasing the required number of binary variables to embed the critical regions, which can be quite large in batch processes.

Assignment constraints. We define a set of binary variables $y_{j,s,t}$ that denote operating state s of an equipment j in time slot t. $y_{j,s,t}$ is equal to 1 if equipment j is occupied by state s in time slot t, and 0 if otherwise. Therefore, we use Eq. 6.16 to dictate the exclusivity of states in an equipment throughout the scheduling time horizon.

$$\sum_{s \in \mathcal{S}} y_{j,s,t} \leq 1, \quad \forall j \in \mathcal{J}, \forall t \in \mathcal{T} \tag{6.16}$$

Similarly, one task can only be executed in one equipment, as given by Eq. 6.17.

$$\sum_{j \in \mathcal{J}} y_{j,s,t} \leq 1, \quad \forall s \in \mathcal{S}, \forall t \in \mathcal{T} \tag{6.17}$$

Continuity constraints. After a task is assigned to an equipment, it has to continue the process in the same equipment.

$$y_{j,s,t+1} \leq y_{j,s,t}, \quad \forall j \in \mathcal{J}, \forall s \in \mathcal{S}, \forall t \in \mathcal{T}, t \neq t_f \tag{6.18}$$

where t_f is the final scheduling time step.

Material balance. At each discretization point, we construct the material balance for every component c to determine their availability, as presented in Eq. 6.19.

$$E_{c,t} = E_{c,t-1} + \sum_{j \in \mathcal{J}} \Delta E_{j,c,t}, \quad \forall c \in \mathcal{C}, \forall t \in \mathcal{T}, t > 0 \tag{6.19}$$

where $E_{c,t}$ denotes the amount of excess material of component c at time t, and $\Delta E_{j,c,t}$ is the generation or consumption term, dictated by the reaction kinetics in the high-fidelity model.

Capacity constraints. The vessel sizes limit the amount of material that can be processed in every task.

$$\sum_{s\in\mathcal{S}} y_{j,s,t}\underline{V} \leq V_{j,t} \leq \sum_{s\in\mathcal{S}} y_{j,s,t}\overline{V}, \quad \forall s \in \mathcal{S}, \forall t \in \mathcal{T} \tag{6.20}$$

where $V_{j,t}$ is a set of continuous variables that describe the volume occupied in equipment j. Note that it is possible to enforce similar constraints on the excess material $E_{c,t}$. However, we assume unlimited intermediate storage (UIS) and neglect such constraints for simplicity.

Quality constraints. These constraints are included to enforce certain quality metrics, such as product purity or target demand, at the end of the batch.

$$\begin{aligned} &x_s^* \leq x_{s,t+1} + \mathcal{M}(w_{s,t}), \quad \forall s \in \mathcal{S}^*, t = 0 \\ &x_s^* \leq x_{s,t+1} + \mathcal{M}\big(1 - (w_{s,t} - w_{s,t+1})\big), \quad \forall s \in \mathcal{S}^*, \forall t \in \mathcal{T}, 0 \leq t \leq t_f \\ &x_s^* \leq x_{s,t+1} + \mathcal{M}\big(1 - w_{s,t}\big), \quad \forall s \in \mathcal{S}^*, t = t_f \end{aligned} \tag{6.21}$$

where superscript "*" denotes the target states, $\mathcal{M}$ is a sufficiently large number for the big-M formulation, and $w_{s,t}$ is defined as a set of linking variables between the scheduling model and the dynamic high fidelity model. The linking variables $w_{s,t}$ are a set of Boolean variables that are enforced to have a "true" value if the task is still in progress via Eq. 6.21, and "false" if otherwise. These variables are linked to the scheduling model as presented by Eq. 6.22.

$$(t+1)w_{s,t} \leq \sum_{j'\in\mathcal{J}} \sum_{t'\in\mathcal{T}} y_{j',s,t'}, \quad \forall s \in \mathcal{S}, \forall t \in \mathcal{T} \tag{6.22}$$

Sequence constraints. In the case that one state should take place only after the completion of another task (e.g. precursors), the priority can be dictated by Eq. 6.23.

$$y_{j,s^+,t} \leq \sum_{t'=0}^{t} y_{j,s^-,t'}, \quad \forall j \in \mathcal{J}, \forall s^- \in \mathcal{S}^-, \forall s^+ \in \mathcal{S}^+, \forall t \in \mathcal{T} \tag{6.23}$$

where superscript "−" denotes the states that should be scheduled earlier than the states labeled by the superscript "+".

Objective functions. Here, we will present two most commonly used objectives in a process schedule, although they can be diversified and tailored to serve different purposes. For makespan minimization, a set of Boolean variables z_t is defined to indicate if the overall process is still in progress.

$$y_{j,s,t} \leq z_t, \quad \forall j \in \mathcal{J}, \forall s \in \mathcal{S} \tag{6.24}$$

Then, the makespan of one batch cycle can be minimized by minimizing the sum of z_t, as presented by Eq. 6.25.

$$\sum_{t\in T} z_t \tag{6.25}$$

Similarly, cost minimization is one of the most common objectives encountered in processes schedules, and can be expressed by Eq. 6.26.

$$\sum_{t \in \mathcal{T}} C^u u_t + \sum_{t \in \mathcal{T}} C^t z_t \tag{6.26}$$

The explicit MPC is expressed by a piecewise affine function, and has two components, namely (i) a set of affine functions that are optimal for the polytopic space CR_n (Eq. 6.2a), and (ii) a set of polytopes that define the space that bound the corresponding affine expression (Eq. 6.2b). Equation 6.2a can be reformulated by using the two main relaxation schemes, namely big-M reformulation and convex hull formulation. These relaxation schemes can be used to embed the mpMPC to the SEN network via a set of binary variables $y_{n,t}^{CR}$.

$$-\mathcal{M}(1 - y_{n,t}^{CR}) \le u_t - K_n\theta_t - r_n \le \mathcal{M}(1 - y_{n,t}^{CR}), \quad \forall n \in NC, \forall t \in \mathcal{T} \tag{6.27a}$$

$$\begin{aligned} &-\mathcal{M}(1 - y_{n,t}^{CR}) \le u_{n,t} - K_n\theta_t - r_n \le \mathcal{M}(1 - y_{n,t}^{CR}), \quad \forall n \in NC, \forall t \in \mathcal{T} \\ &u_t = \sum_{n \in NC} u_{n,t}, \quad \forall t \in \mathcal{T} \end{aligned} \tag{6.27b}$$

where Eq. 6.27a represents the big-M reformulation and Eq. 6.27b represents the convex hull reformulation for the optimal control law. We also dictate that at most one critical region can be selected at a given time throughout the scheduling horizon, as given by Eq. 6.28.

$$\sum_{n \in NC} y_{n,t}^{CR} \le 1, \quad \forall t \in \mathcal{T} \tag{6.28}$$

Selection of the critical region strictly depends on the feasibility of the parameter set $\boldsymbol{\theta}_t$ at time t according to Eq. 6.2b. Therefore, we can simply relax the disjoint polytopes by Eq. 6.29.

$$A_n^{CR}\theta_t - b_n^{CR} \le \mathcal{M}(1 - y_{n,t}^{CR}), \quad \forall n \in NC, \forall t \in \mathcal{T} \tag{6.29}$$

Note that both the big-M and convex hull reformulation schemes require a binary variable for every critical region and for each time step throughout the scheduling horizon. Consequently, the computational complexity of the MIDO problem grow exponentially as the number of critical regions of the explicit optimal control law increase. The states of a batch process are inherently time-varying and hence, the MPC scheme of a batch process requires longer output and control horizons, and larger bounds on the variables compared to a typical continuous process. The combinatorial nature of the increased number of variables and constraints of the MPC problem results into an exponential increase in the number of critical regions in its explicit solution. Therefore, employing the big-M and convex hull reformulation techniques become impractical due to the number of the $y_{n,t}^{CR}$ variables, especially for the batch processes. Herein, we present an efficient modeling technique with significantly less binary variables using the base-2 numeral system. The goal of this technique is to represent each critical region in a time step with a unique combination of a set of binary variables, $\bar{y}_{n_2,t}^{CR}$.

Let n_2 denote the nth critical region in the base-2 numeral system (i.e. $n_2 = n$). We treat the digits of n_2 as an array of binary parameters, denoted by β_{n_2}. Therefore, Eqs. 6.27 and 6.29 can be relaxed with the unique combinations of a set of binary variables y_i as presented by Eq. 6.30 for the big-M relaxation and Eq. 6.31 for the convex hull relaxation instead.

$$-\mathcal{M}\Bigg(\sum_{i\in\{m|\beta_{n_2,m}=0\}} \bar{y}_{i,t}^{CR} + \sum_{i\in\{m|\beta_{n_2,m}=1\}} (1-\bar{y}_{i,t}^{CR})\Bigg) \le u_t - K_{n_2}\theta_t - r_{n_2}, \quad \forall n_2 \in NC_2, \forall t \in \mathcal{T} \tag{6.30a}$$

$$u_t - K_{n_2}\theta_t - r_{n_2} \le \mathcal{M}\Bigg(\sum_{i\in\{m|\beta_{n_2,m}=0\}} \bar{y}_{i,t}^{CR} + \sum_{i\in\{m|\beta_{n_2,m}=1\}} (1-\bar{y}_{i,t}^{CR})\Bigg), \quad \forall n_2 \in NC_2, \forall t \in \mathcal{T} \tag{6.30b}$$

$$-\mathcal{M}\Bigg(\sum_{i\in\{m|\beta_{n_2,m}=0\}} \bar{y}_{i,t}^{CR} + \sum_{i\in\{m|\beta_{n_2,m}=1\}} (1-y_{i,t}^{CR})\Bigg) \le u_{n_2,t} - K_{n_2}\theta_t - r_{n_2}, \quad \forall n_2 \in NC_2, \forall t \in \mathcal{T} \tag{6.31a}$$

$$u_{n_2,t} - K_{n_2}\theta_t - r_{n_2} \le \mathcal{M}\Bigg(\sum_{i\in\{m|\beta_{n_2,m}=0\}} \bar{y}_{i,t}^{CR} + \sum_{i\in\{m|\beta_{n_2,m}=1\}} (1-\bar{y}_{i,t}^{CR})\Bigg), \quad \forall n_2 \in NC_2, \forall t \in \mathcal{T} \tag{6.31b}$$

$$u_t = \sum_{n_2\in NC_2} u_{n_2,t}, \quad \forall t \in \mathcal{T} \tag{6.31c}$$

Note that we do not enforce Eq. 6.28 in the base-2 numeral system as any feasible combination of the binary variables $\bar{y}_{n_2,t}^{CR}$ yield a unique optimal control law. The feasibility of the control laws in closed loop is analogously satisfied by Eq. 6.32.

$$A_{n_2}^{CR}\theta_t - b_{n_2}^{CR} \le \mathcal{M}\Bigg(\sum_{i\in\{m|\beta_{n_2,m}=0\}} \bar{y}_{i,t}^{CR} + \sum_{i\in\{m|\beta_{n_2,m}=1\}} (1-\bar{y}_{i,t}^{CR})\Bigg), \quad \forall n_2 \in NC_2, \forall t \in \mathcal{T} \tag{6.32}$$

Therefore, using Eqs. 6.30 or 6.27b along with Eq. 6.32 provides an exact integration of the mpMPC into the MIDO formulation. If the number of critical regions n is greater than the number of binary combinations (i.e. $2^{\lceil \log_2 n \rceil} > n$), then we can use Eq. 6.33 to eliminate the infeasible combinations.

$$\sum_{i\in\{m|\beta_{n_2,m}=1\}} \bar{y}_{i,t}^{CR} - \sum_{i\in\{m|\beta_{n_2,m}=0\}} \bar{y}_{i,t}^{CR} \le \mid m|\beta_{n_2,m} = 1 \mid -1, \quad t \in \mathcal{T} \tag{6.33}$$

where $|\cdot|$ denotes the cardinality operator.

Highfidelity dynamic model. We consider a system of three sets of reactions taking place in two reactors, where both reactors are capable of processing the available tasks. The stoichiometry of the reactions is presented as follows.

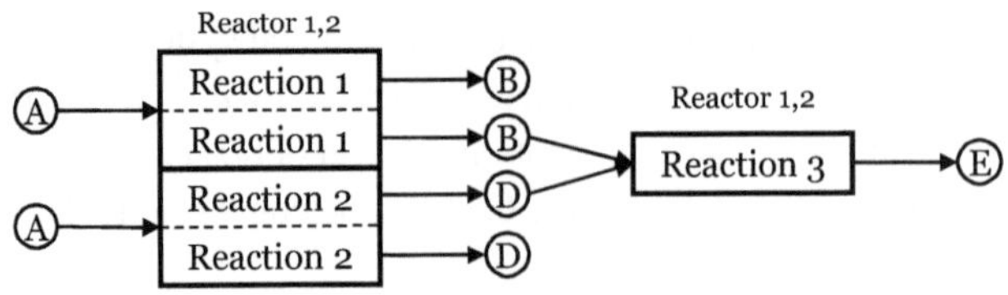

Fig. 6.12 SEN representation of the multipurpose batch process example

$$\text{Reaction 1: } A \underset{k_{-1}}{\overset{k_1}{\rightleftharpoons}} B \overset{k_2}{\rightarrow} C$$

$$\text{Reaction 2: } A \overset{k_3}{\rightarrow} D$$

$$\text{Reaction 3: } B + D \overset{k_4}{\rightarrow} E$$

where the valuable products are B, D, and E. Therefore, the operator has the degree of freedom to select the most convenient task that delivers the requirements of the desired objective at a given time. We employ the SEN framework to determine the process schedule over a given horizon. The SEN representation of the process is shown in Fig. 6.12. The mathematical model presented in Eqs. 6.34–6.36 are used to simulate the dynamic behavior of the system, and the parameters of the reactions are provided in Table 6.3.

$$\frac{1}{V}\frac{dN_c}{dt} = \sum_{r \in \mathcal{R}} s_{c,r} r_r, \quad \forall c \in \mathcal{C} \tag{6.34a}$$

$$\frac{dT}{dt} = \frac{-\sum_{r \in \mathcal{R}} r_r \Delta H_r + Q/V}{\rho c_p} \tag{6.34b}$$

where V is the working volume, N_c is the amount of component c, $s_{c,r}$ is the stoichiometric coefficient of product c in reaction r, r_r is the rate of reaction of r, T is the temperature, ΔH_r is the reaction enthalpy, Q is the heat input, ρ is the mixture molar density, and c_p is the specific heat capacity.

Table 6.3 Parameters of the batch process

Reaction	ΔH_r [kJ/mol]	k_r	$E_{A,r}$ [kJ/mol.K]
k_1	13.8	5.0×10^6 h^{-1}	50.8
k_{-1}	−13.8	0.1×10^6 h^{-1}	37.0
k_2	−2.0	0.5×10^6 h^{-1}	46.0
k_3	9.8	5.0×10^3 h^{-1}	90
k_4	−10	0.5 m^3 kmol^{-1}h^{-1}	40.7

Table 6.4 mpMPC tuning parameters for the reactions in the multipurpose batch process example

Parameter	Reaction 1	Reaction 2	Reaction 3
OH	$\{1, 2, 3, 4\}$	$\{1, 2, 3, 4\}$	$\{1, 2, 3, 4\}$
CH	$\{1, 2, 3\}$	$\{1, 2, 3\}$	$\{1, 2, 3\}$
$QR_k, \forall k \in OH$	$\begin{bmatrix} 0 & 0 \\ 0 & 10^4 \end{bmatrix}$	$\begin{bmatrix} 0 & 0 \\ 0 & 10^3 \end{bmatrix}$	$\begin{bmatrix} 0 & 0 \\ 0 & 4 \times 10^3 \end{bmatrix}$
$R1_k, \forall k \in OH$	1.0	1.0	1.0

The rate expressions for all three reactions are given by Eq. 6.35.

$$r_r = k_r \exp\left(-\frac{E_{A,r}}{RT}\right) \prod_{c \in rxn_r} \frac{N_c}{V}, \quad \forall r \in \mathcal{R} \tag{6.35}$$

where k_r is the pre-exponential term of the Arrhenius equation, $E_{A,r}$ is the activation energy, R is the ideal gas constant, and rxn_r is the set of components that take place in reaction r. Note that all reactions are assumed to be first order with respect to the reactants.

Lastly, the density and the heat capacity of the mixture are determined by assuming ideal conditions.

$$\begin{aligned} \rho &= \frac{\sum_{c \in \mathcal{C}} N_c}{V} \\ c_p &= \frac{\sum_{c \in \mathcal{C}} c_{p,c} N_c}{\sum_{c \in \mathcal{C}} N_c} \end{aligned} \tag{6.36}$$

The objective of this problem is to maximize the profit, which accounts for the revenue from selling the products, the operating cost for the heat supply and raw material purchases, and the investment cost due to the sizing of the reactors. The design decisions are particularly included in this case study, as they directly conflict with the operating decisions in terms of the optimality and feasibility of the problem.

Model approximation and mpMPC development. We follow the procedure to develop the explicit MPC strategies described in the earlier examples. The approximate state space models for these systems are provided in Appendix D.2.2. The tuning parameters of the control strategies are presented in Table 6.4. The formulated mpMPC is solved using the POP toolbox in the MATLAB environment [376].[1]

[1] The explicit MPC solutions can be downloaded from http://paroc.tamu.edu/.

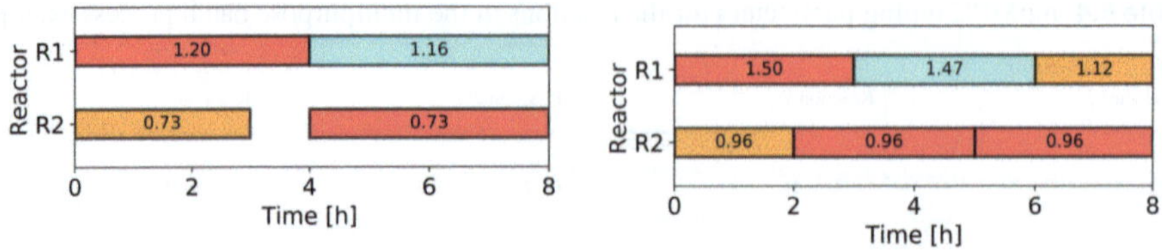

(a) Process schedule with the MPC dynamics included. The reactor sizes are $V_{R1} = 1.20m^3$ and $V_{R2} = 0.73m^3$, respectively.

(b) Process schedule without the MPC dynamics. The reactor sizes are $V_{R1} = 1.50m^3$ and $V_{R2} = 0.96m^3$, respectively.

Fig. 6.13 Process schedule with and without considering the MPC dynamics. The colors red, orange, and cyan represent the production of B, D, and E, respectively

Dynamic optimization. In this example, the schedule is designed over a horizon of 8 h. We assume that the processing time for the separation of the product of interest from the unreacted raw materials and by-products is negligible. The integrated MIDO problem is reformulated as an MINLP by orthogonal collocation on finite elements with 24 finite elements and 3 collocation points over each finite element. The reformulated MINLP is solved with GAMS/BARON [538] with a 15 min limit on the solution time. The resulting process schedule is demonstrated on a Gantt chart in Fig. 6.13a. By the end of the scheduling horizon, the targeted inventory is 0.28 kmol for B and 0.39 kmol for E, while no excess D is produced. For reference, the same problem is solved without accounting for the dynamics introduced by the MPC, which is presented in Fig. 6.13b. Here, the scheduler aims to produce 0.76 kmol B, 1.02 kmol D, and 0.49 kmol E by the end of the horizon. Notice that the targeted inventory levels are in fact lower when the MPC dynamics are included in the integrated problem. Acquiring "worse" solutions with smaller profit margins with the proposed approach is an expected outcome since the problem without the MPC dynamics is an underestimator[2] of the completely integrated problem. The most imperative contribution of the proposed approach is to provide certificates of operability for the calculated optimal trajectory under the simultaneously determined process design. The benefit of having such certificates can be observed in Fig. 6.14, where the optimal trajectories are determined based on closed-loop and open loop dynamic optimization formulations for three distinct tasks. The proposed approach bridges the gap between the optimal trajectory and its realization in closed-loop by distinguishing the desired path from the set points that need to be passed on to the controller. Following these trajectories, the proposed scheduling and control scheme achieves the targeted inventory levels by producing 0.28 kmol B and 0.38 kmol E at the end of the horizon. However in the reference case, the controllers fail quite often due to infeasible parameter realizations. The steep changes in set points impose unrealistic trajectories for the controller, which cannot satisfy the terminal constraints. Therefore, the closed-loop

[2] Underestimator is used in the direction of a conventional minimization problem.

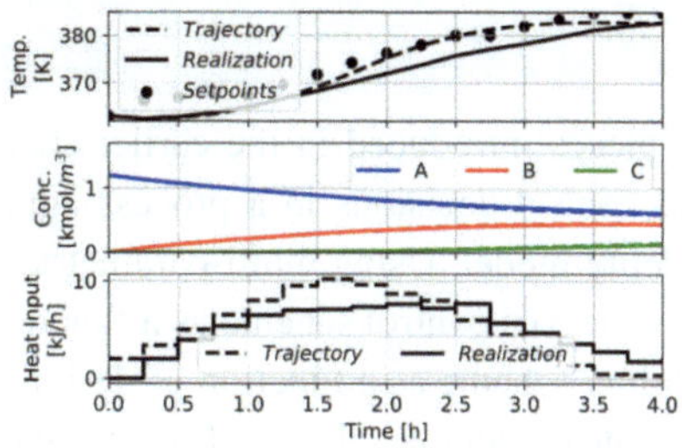

(a) Production of B based on closed-loop optimization.

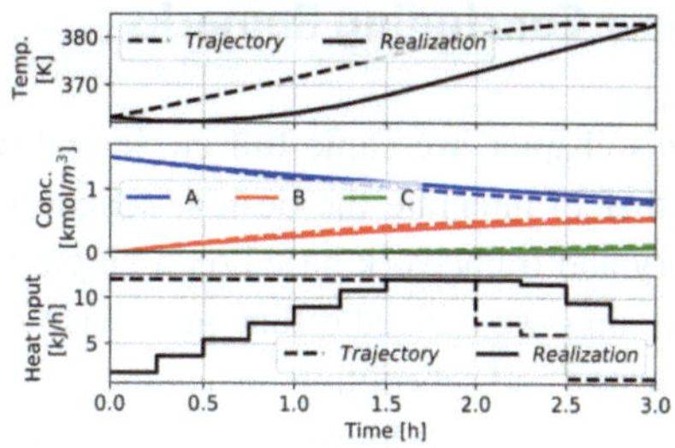

(b) Production of B based on open loop optimization.

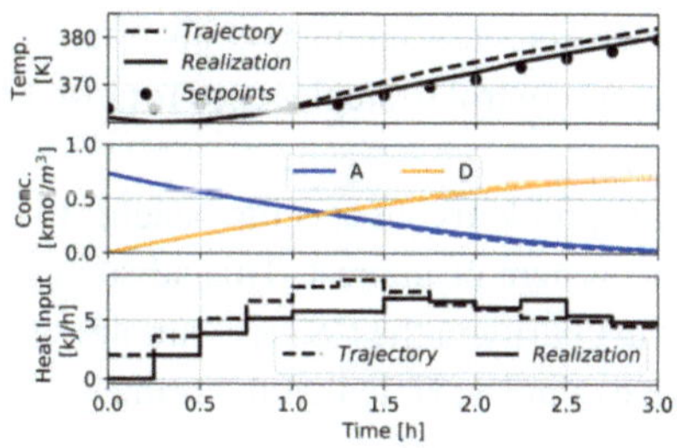

(c) Production of D based on closed-loop optimization.

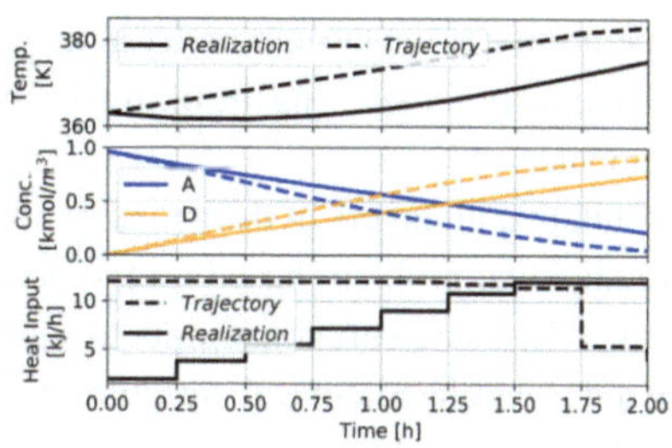

(d) Production of D based on open loop optimization.

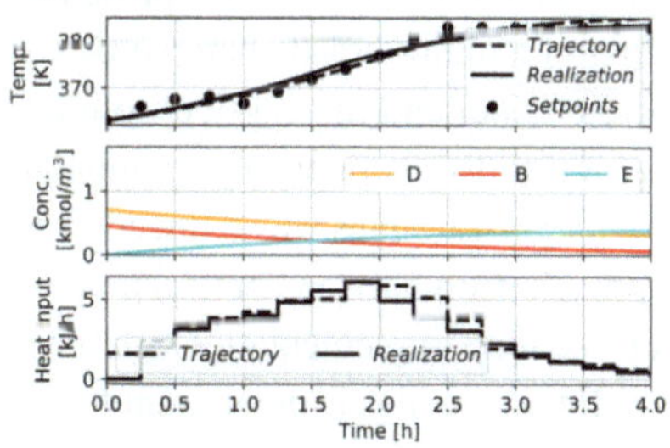

(e) Production of E based on closed-loop optimization.

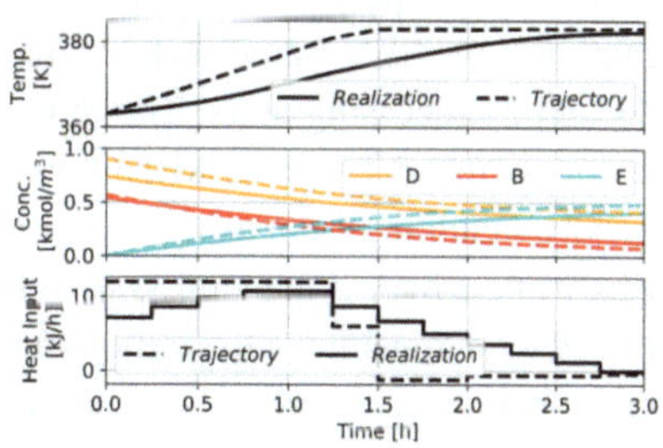

(f) Production of E based on open loop optimization.

Fig. 6.14 Closed loop implementation of the optimal input-output trajectories for three different tasks against the closed-loop and open loop dynamic optimization formulations. Figures on the left hand side (Fig. 6.14a, c, and e) show the optimal closed-loop trajectories, set-points that are passed on to the MPC, and the closed-loop realizations. Figures on the right hand side (Fig. 6.14b, d, and f) only show the optimal trajectories and their closed-loop realizations since the trajectories are used as the set points

realizations in Fig. 6.14b, d, and f are simulated without enforcing the terminal constraints in the MPC. Due to the mismatch between the set points and the closed-loop realization, the targeted production cannot be achieved in the dedicated time interval. In other words, the schedule cannot be satisfied due to the delay in delivering the intended amounts.

6.5 Concluding Remarks

In this chapter, we came full circle and put the pieces introduced in the earlier chapters together to integrate the design, scheduling, and control decisions in a process agnostic theory and framework based on a single high fidelity model. Using multi-parametric programming, we derived offline piecewise strategies for (i) a control scheme as a function of the design and scheduling decisions, (ii) a scheduling scheme as a function of design, and aware of the closed-loop dynamics through a surrogate model formulation. The offline maps of strategies allowed for a direct implementation in a MIDO formulation for design optimization. The proposed framework was able to determine the process design that guarantees the operability of the system under a range of bounded process and market uncertainties by simultaneously considering the optimal scheduling and control strategies used in closed-loop implementation.

Postulating all layers of decisions as optimization problems has specific benefits to tailor each individual problem based on the needs of the system of interest. This advantage was illustrated by using soft constraints to satisfy product purity in the CSTR examples, and by using a decentralized control structure in the CHP examples. Note that the framework was applied on both problems without appealing to further modifications.

We further demonstrated that the framework is also applicable on batch processes where the scheduling problem requires more sophisticated mathematical representations. We employed the SEN framework for the scheduling of a multipurpose batch process with 3 tasks and 2 processing units. The dynamic process model and the explicit optimal control law was embedded in the MIDO problem via logical disjunctions. The computational complexity stemming from using a large number of binary variables was reduced exponentially by using a base-2 numeral system based approach. Comparisons between open loop and closed loop dynamic optimizations and their closed loop implementations on the actual process dynamics revealed that accounting for the MPC dynamics is paramount to provide certificates of operability.

The work presented in this book offers an theoretical improvement for the Process Systems Engineering community to simultaneously address process design and operational optimization decisions. Further improvement on this approach should be expected with the advances on two fronts: (i) theoretical developments in deriving an exact multi-parametric solution for robust MPC, and (ii) computational developments to solve large scale dynamic optimization and multi-parametric programming. The former is essential to push the limits of the representation accuracy of operational closed loop decisions in the overarching optimization layer. The approximation steps in the framework is the major root of mismatch between the real process dynamics and the decision making optimization problems. Although this mismatch is currently accounted for as right hand side parameters in the control and scheduling problems, left hand side representation of these parameters will increase the model accuracy as it will capture the nonlinear dynamics in the process. The latter point is the practical point to improve the tractability of the derived integrated problems. Dynamic optimization is already a challenging class of problems due to the infinite dimensional variables. In these particular integrated problems, the number of binary variables increase with the time horizon. Therefore, a realistic and representatively long time horizon will require the use a significant number of binary variables, rendering the problem intractable. However, tailored MIDO algorithms can be developed to fathom infeasible and suboptimal integer combinations via exploiting the structure of the piecewise affine nature of the parametric solution space. We expect that further advances in these fields should bring us one step closer to seamless integration of process design and operational optimization decisions, both in theory and real life application.

Supporting Information for Chap. 2

A

A.1 Proof that Master Problems Eqs. 2.7 and 2.12 are Equivalent

In order to show that master problems Eqs. 2.7 and 2.12 are equivalent, equivalence needs to be established between the Lagrange functions $\bar{\mathcal{L}}^k$ in Eq. 2.6 and $\mathcal{L}^k$ in Eq. 2.11. This can be achieved as follows. Firstly, consider the optimality conditions that must be satisfied by the continuous variables $\boldsymbol{\omega}$ in Eq. 2.9. These are determined by finding a stationary point for the Lagrange function $\bar{\mathcal{L}}^k$ in Eq. 2.10, i.e.:

$$
\begin{aligned}
0 = & \left(\frac{\partial J}{\partial \boldsymbol{\omega}}\right)^T\Bigg|_{\boldsymbol{\omega}^k} + \int_{t_0}^{t_f} \left(\frac{\partial \mathbf{h}_d}{\partial \boldsymbol{\omega}}\right)^T\Bigg|_{\boldsymbol{\omega}^k} \cdot \boldsymbol{\nu}_d^k dt + \int_{t_0}^{t_f} \left(\frac{\partial \mathbf{h}_a}{\partial \boldsymbol{\omega}}\right)^T\Bigg|_{\boldsymbol{\omega}^k} \cdot \boldsymbol{\nu}_a^k dt \\
& + \left(\frac{\partial \mathbf{h}_d}{\partial \boldsymbol{\omega}}\right)^T\Bigg|_{t_f,\boldsymbol{\omega}^k} \cdot \boldsymbol{\xi}_{d,f}^k dt + \left(\frac{\partial \mathbf{h}_a}{\partial \boldsymbol{\omega}}\right)^T\Bigg|_{t_f,\boldsymbol{\omega}^k} \cdot \boldsymbol{\xi}_{a,f}^k dt \\
& + \left(\frac{\partial \mathbf{h}_d}{\partial \boldsymbol{\omega}}\right)^T\Bigg|_{t_0,\boldsymbol{\omega}^k} \cdot \boldsymbol{\xi}_{d,0}^k dt + \left(\frac{\partial \mathbf{h}_a}{\partial \boldsymbol{\omega}}\right)^T\Bigg|_{t_0,\boldsymbol{\omega}^k} \cdot \boldsymbol{\xi}_{a,0}^k dt + \left(\frac{\partial \mathbf{h}_0}{\partial \boldsymbol{\omega}}\right)^T\Bigg|_{\boldsymbol{\omega}^k} \cdot \boldsymbol{\kappa}^k dt \\
& + \left(\frac{\partial \mathbf{h}_p}{\partial \boldsymbol{\omega}}\right)^T\Bigg|_{\boldsymbol{\omega}^k} \cdot \boldsymbol{\mu}_p^k dt + \left(\frac{\partial \mathbf{g}_p}{\partial \boldsymbol{\omega}}\right)^T\Bigg|_{\boldsymbol{\omega}^k} \cdot \boldsymbol{\lambda}_p^k dt + \left(\frac{\partial \mathbf{h}_q}{\partial \boldsymbol{\omega}}\right)^T\Bigg|_{\boldsymbol{\omega}^k} \cdot \boldsymbol{\mu}_q^k dt \\
& + \left(\frac{\partial \mathbf{g}_q}{\partial \boldsymbol{\omega}}\right)^T\Bigg|_{\boldsymbol{\omega}^k} \cdot \boldsymbol{\lambda}_q^k dt + \mathbf{I} \cdot \boldsymbol{\zeta}^k
\end{aligned}
\tag{A.1}
$$

Note that in Eq. A.1, all of the Jacobian matrices appear in transposed form. Thus, $(\partial J/\partial \boldsymbol{\omega})$ is a $(1 \times n_{\boldsymbol{\omega}})$ vector, $(\partial \mathbf{h}_d)/\partial \boldsymbol{\omega}$ is a $(n_{\mathbf{x}_d} \times n_{\boldsymbol{\omega}})$ matrix, and so on, where $n_{\boldsymbol{\omega}}$ and $n_{\mathbf{x}_d}$ are the numbers of binary and differential variables, respectively.

B. Burnak et al., *Integrated Process Design and Operational Optimization via Multi-parametric Programming*, Synthesis Lectures on Engineering, Science, and Technology, https://doi.org/10.1007/978-3-032-10181-5

Substituting the expression for ζ^k given by Eq. A.1 into Eq. 2.11 then gives:

$$\begin{aligned}
\bar{\mathcal{L}}^k(\hat{\mathbf{u}}^k\hat{\mathbf{d}}^k, \bar{\omega}) = & \mathbf{J}\big(\hat{\dot{\mathbf{x}}}_d^k(t_f), \hat{\mathbf{x}}_d^k(t_f), \hat{\mathbf{x}}_a^k(t_f), \hat{\mathbf{u}}^k(t_f), \hat{\mathbf{d}}^k(t_f), \boldsymbol{\omega}^k, t_f\big) \\
& + \Bigg[\left(\frac{\partial \mathbf{J}}{\partial \boldsymbol{\omega}}\right)\bigg|_{\boldsymbol{\omega}^k} + \int_{t_0}^{t_f} \boldsymbol{\nu}_d^{k\,T}\left(\frac{\partial \mathbf{h}_d}{\partial \boldsymbol{\omega}}\right)\bigg|_{\boldsymbol{\omega}^k} dt + \int_{t_0}^{t_f} \boldsymbol{\nu}_a^{k\,T}\left(\frac{\partial \mathbf{h}_a}{\partial \boldsymbol{\omega}}\right)\bigg|_{\boldsymbol{\omega}^k} dt \\
& + \boldsymbol{\xi}_{d,f}^{k\;T}\left(\frac{\partial \mathbf{h}_d}{\partial \boldsymbol{\omega}}\right)\bigg|_{t_f,\boldsymbol{\omega}^k} + \boldsymbol{\xi}_{a,f}^{k\;T}\left(\frac{\partial \mathbf{h}_a}{\partial \boldsymbol{\omega}}\right)\bigg|_{t_f,\boldsymbol{\omega}^k} \\
& + \boldsymbol{\xi}_{d,0}^{k\;T}\left(\frac{\partial \mathbf{h}_d}{\partial \boldsymbol{\omega}}\right)\bigg|_{t_0,\boldsymbol{\omega}^k} + \boldsymbol{\xi}_{a,0}^{k\;T}\left(\frac{\partial \mathbf{h}_a}{\partial \boldsymbol{\omega}}\right)\bigg|_{t_0,\boldsymbol{\omega}^k} + \boldsymbol{\kappa}^{k\,T}\left(\frac{\partial \mathbf{h}_0}{\partial \boldsymbol{\omega}}\right)\bigg|_{\boldsymbol{\omega}^k} \\
& + \boldsymbol{\mu}_p^{k\,T}\left(\frac{\partial \mathbf{h}_p}{\partial \boldsymbol{\omega}}\right)\bigg|_{\boldsymbol{\omega}^k} + \boldsymbol{\lambda}_p^{k\,T}\left(\frac{\partial \mathbf{g}_p}{\partial \boldsymbol{\omega}}\right)\bigg|_{\boldsymbol{\omega}^k} \\
& + \boldsymbol{\mu}_q^{k\,T}\left(\frac{\partial \mathbf{h}_q}{\partial \boldsymbol{\omega}}\right)\bigg|_{\boldsymbol{\omega}^k} + \boldsymbol{\lambda}_q^{k\,T}\left(\frac{\partial \mathbf{g}_q}{\partial \boldsymbol{\omega}}\right)\bigg|_{\boldsymbol{\omega}^k}\Bigg](\bar{\omega} - \boldsymbol{\omega}^k)
\end{aligned} \tag{A.2}$$

Now consider the original Lagrange function $\mathcal{L}^k$, Eq. 2.6, used in the master problems of the algorithms of Mohideen et al. [264] and Schweiger and Floudas [261]. For a system that is linear in y (as those algorithms require), the objective function term in Eq. 2.6 can be decomposed as:

$$\begin{aligned}
& \mathbf{J}\big(\hat{\dot{\mathbf{x}}}_d^k(t_f), \hat{\mathbf{x}}_d^k(t_f), \hat{\mathbf{x}}_a^k(t_f), \hat{\mathbf{u}}^k(t_f), \hat{\mathbf{d}}^k(t_f), \boldsymbol{\omega}, t_f\big) \\
& = \mathbf{J}_1\big(\hat{\dot{\mathbf{x}}}_d^k(t_f), \hat{\mathbf{x}}_d^k(t_f), \hat{\mathbf{x}}_a^k(t_f), \hat{\mathbf{u}}^k(t_f), \hat{\mathbf{d}}^k(t_f), t_f\big) \\
& \quad + \mathbf{J}_2\big(\hat{\dot{\mathbf{x}}}_d^k(t_f), \hat{\mathbf{x}}_d^k(t_f), \hat{\mathbf{x}}_a^k(t_f), \hat{\mathbf{u}}^k(t_f), \hat{\mathbf{d}}^k(t_f), t_f\big)\boldsymbol{\omega}
\end{aligned} \tag{A.3}$$

Similarly, the objective function evaluated at $\boldsymbol{\omega} = \boldsymbol{\omega}^k$ is:

$$\begin{aligned}
& \mathbf{J}\big(\hat{\dot{\mathbf{x}}}_d^k(t_f), \hat{\mathbf{x}}_d^k(t_f), \hat{\mathbf{x}}_a^k(t_f), \hat{\mathbf{u}}^k(t_f), \hat{\mathbf{d}}^k(t_f), \boldsymbol{\omega}^k, t_f\big) \\
& = \mathbf{J}_1\big(\hat{\dot{\mathbf{x}}}_d^k(t_f), \hat{\mathbf{x}}_d^k(t_f), \hat{\mathbf{x}}_a^k(t_f), \hat{\mathbf{u}}^k(t_f), \hat{\mathbf{d}}^k(t_f), t_f\big) \\
& \quad + \mathbf{J}_2\big(\hat{\dot{\mathbf{x}}}_d^k(t_f), \hat{\mathbf{x}}_d^k(t_f), \hat{\mathbf{x}}_a^k(t_f), \hat{\mathbf{u}}^k(t_f), \hat{\mathbf{d}}^k(t_f), t_f\big)\boldsymbol{\omega}^k
\end{aligned} \tag{A.4}$$

Eliminating $\mathbf{J}_1$ from Eqs. A.3 and A.4 gives:

$$\begin{aligned}
& \mathbf{J}\big(\hat{\dot{\mathbf{x}}}_d^k(t_f), \hat{\mathbf{x}}_d^k(t_f), \hat{\mathbf{x}}_a^k(t_f), \hat{\mathbf{u}}^k(t_f), \hat{\mathbf{d}}^k(t_f), \boldsymbol{\omega}, t_f\big) \\
& = \mathbf{J}\big(\hat{\dot{\mathbf{x}}}_d^k(t_f), \hat{\mathbf{x}}_d^k(t_f), \hat{\mathbf{x}}_a^k(t_f), \hat{\mathbf{u}}^k(t_f), \hat{\mathbf{d}}^k(t_f), \boldsymbol{\omega}^k, t_f\big) \\
& \quad + \mathbf{J}_2\big(\hat{\dot{\mathbf{x}}}_d^k(t_f), \hat{\mathbf{x}}_d^k(t_f), \hat{\mathbf{x}}_a^k(t_f), \hat{\mathbf{u}}^k(t_f), \hat{\mathbf{d}}^k(t_f), t_f\big)(\boldsymbol{\omega} - \boldsymbol{\omega}^k)
\end{aligned} \tag{A.5}$$

Now $\mathbf{J}_2$ is simply equal to $(\partial \mathbf{J}/\partial \boldsymbol{\omega})$ and so Eq. A.5 can be written as:

$$\begin{aligned}&\mathbf{J}\big(\hat{\dot{\mathbf{x}}}_d^k(t_f), \hat{\mathbf{x}}_d^k(t_f), \hat{\mathbf{x}}_a^k(t_f), \hat{\mathbf{u}}^k(t_f), \hat{\mathbf{d}}^k(t_f), \boldsymbol{\omega}, t_f\big)\\ &= \mathbf{J}\big(\hat{\dot{\mathbf{x}}}_d^k(t_f), \hat{\mathbf{x}}_d^k(t_f), \hat{\mathbf{x}}_a^k(t_f), \hat{\mathbf{u}}^k(t_f), \hat{\mathbf{d}}^k(t_f), \boldsymbol{\omega}^k, t_f\big) + \left(\frac{\partial \mathbf{J}}{\partial \boldsymbol{\omega}}\right)\bigg|_{\boldsymbol{\omega}^k} (\boldsymbol{\omega} - \boldsymbol{\omega}^k)\end{aligned} \tag{A.6}$$

In an analogous manner to that above, the terms within the Lagrange function Eq. 2.6 that relate to the equality constraints can be written as:

$$\mathbf{h}_{\sqcup}(\ldots, \boldsymbol{\omega}, \ldots) = \mathbf{h}_{\sqcup}(\ldots, \boldsymbol{\omega}^k, \ldots) + \left(\frac{\partial \mathbf{h}_{\sqcup}}{\partial \boldsymbol{\omega}}\right)\bigg|_{\boldsymbol{\omega}^k} (\boldsymbol{\omega} - \boldsymbol{\omega}^k), \quad \sqcup = [d, a, 0, p, q] \tag{A.7}$$

with the added simplification that $\mathbf{h}_{\sqcup}(\ldots, \boldsymbol{\omega}^k, \ldots) = 0$ at the primal solution, i.e.

$$\mathbf{h}_{\sqcup}(\ldots, \boldsymbol{\omega}, \ldots) = \left(\frac{\partial \mathbf{h}_{\sqcup}}{\partial \boldsymbol{\omega}}\right)\bigg|_{\boldsymbol{\omega}^k} (\boldsymbol{\omega} - \boldsymbol{\omega}^k), \quad \sqcup = [d, a, 0, p, q] \tag{A.8}$$

Similarly, the terms in Eq. 2.6 that pertain to the inequality constraints become:

$$\boldsymbol{\lambda}_{\sqcup}^{k\,T} \mathbf{g}_{\sqcup}(\ldots, \boldsymbol{\omega}, \ldots) = \boldsymbol{\lambda}_{\sqcup}^{k\,T} \mathbf{g}_{\sqcup}(\ldots, \boldsymbol{\omega}^k, \ldots) + \left(\frac{\partial \mathbf{g}_{\sqcup}}{\partial \boldsymbol{\omega}}\right)\bigg|_{\boldsymbol{\omega}^k} (\boldsymbol{\omega} - \boldsymbol{\omega}^k), \quad \sqcup = [p, q] \tag{A.9}$$

with the simplification that $\boldsymbol{\lambda}_{\sqcup}^{k\,T} \mathbf{g}_{\sqcup}(\ldots, \boldsymbol{\omega}^k, \ldots) = 0$ due to the complementarity optimality conditions of the primal problem. Hence:

$$\boldsymbol{\lambda}_{\sqcup}^{k\,T} \mathbf{g}_{\sqcup}(\ldots, \boldsymbol{\omega}, \ldots) = \left(\frac{\partial \mathbf{g}_{\sqcup}}{\partial \boldsymbol{\omega}}\right)\bigg|_{\boldsymbol{\omega}^k} (\boldsymbol{\omega} - \boldsymbol{\omega}^k), \quad \sqcup = [p, q] \tag{A.10}$$

Substituting Eqs. A.6, A.8 A.10 in Eq. 2.6 gives:

$$\begin{aligned}\bar{\mathcal{L}}^k(\hat{\mathbf{u}}^k\hat{\mathbf{d}}^k, \bar{\omega}) = &\,\mathbf{J}\big(\hat{\dot{\mathbf{x}}}_d^k(t_f), \hat{\mathbf{x}}_d^k(t_f), \hat{\mathbf{x}}_a^k(t_f), \hat{\mathbf{u}}^k(t_f), \hat{\mathbf{d}}^k(t_f), \boldsymbol{\omega}^k, t_f\big)\\ &+ \Bigg[\left(\frac{\partial \mathbf{J}}{\partial \boldsymbol{\omega}}\right)\bigg|_{\boldsymbol{\omega}^k} + \int_{t_0}^{t_f} \boldsymbol{\nu}_d^{k\,T} \left(\frac{\partial \mathbf{h}_d}{\partial \boldsymbol{\omega}}\right)\bigg|_{\boldsymbol{\omega}^k} dt + \int_{t_0}^{t_f} \boldsymbol{\nu}_a^{k\,T} \left(\frac{\partial \mathbf{h}_a}{\partial \boldsymbol{\omega}}\right)\bigg|_{\boldsymbol{\omega}^k} dt\\ &+ \boldsymbol{\xi}_{d,f}^{k\ T} \left(\frac{\partial \mathbf{h}_d}{\partial \boldsymbol{\omega}}\right)\bigg|_{t_f, \boldsymbol{\omega}^k} + \boldsymbol{\xi}_{a,f}^{k\ T} \left(\frac{\partial \mathbf{h}_a}{\partial \boldsymbol{\omega}}\right)\bigg|_{t_f, \boldsymbol{\omega}^k}\\ &+ \boldsymbol{\xi}_{d,0}^{k\ T} \left(\frac{\partial \mathbf{h}_d}{\partial \boldsymbol{\omega}}\right)\bigg|_{t_0, \boldsymbol{\omega}^k} + \boldsymbol{\xi}_{a,0}^{k\ T} \left(\frac{\partial \mathbf{h}_a}{\partial \boldsymbol{\omega}}\right)\bigg|_{t_0, \boldsymbol{\omega}^k} + \boldsymbol{\kappa}^{k\,T} \left(\frac{\partial \mathbf{h}_0}{\partial \boldsymbol{\omega}}\right)\bigg|_{\boldsymbol{\omega}^k}\\ &+ \boldsymbol{\mu}_p^{k\,T} \left(\frac{\partial \mathbf{h}_p}{\partial \boldsymbol{\omega}}\right)\bigg|_{\boldsymbol{\omega}^k} + \boldsymbol{\lambda}_p^{k\,T} \left(\frac{\partial \mathbf{g}_p}{\partial \boldsymbol{\omega}}\right)\bigg|_{\boldsymbol{\omega}^k}\\ &+ \boldsymbol{\mu}_q^{k\,T} \left(\frac{\partial \mathbf{h}_q}{\partial \boldsymbol{\omega}}\right)\bigg|_{\boldsymbol{\omega}^k} + \boldsymbol{\lambda}_q^{k\,T} \left(\frac{\partial \mathbf{g}_q}{\partial \boldsymbol{\omega}}\right)\bigg|_{\boldsymbol{\omega}^k}\Bigg](\bar{\boldsymbol{\omega}} - \boldsymbol{\omega}^k)\end{aligned} \tag{A.11}$$

which is identical to Eq. A.2 as required.

A.2 Demonstrating the Equivalence of GBD and OA the Example Problem

In Step 3 of Algorithm I/II, a dynamic optimization problem is solved. During the course of this, the optimizer "sees" an NLP formed from the elimination of the differential and algebraic variables during an integration phase. For the given example, this NLP takes the form:

$$\begin{aligned} \min_{V,R,\mathbf{yf}} \quad & ISE(V,T,\mathbf{yf}) \\ s.t. \quad & g_1(V,R,\mathbf{yf}) \leq 0 \\ & g_2(V,R,\mathbf{yf}) \leq 0 \\ & \mathbf{yf} - \bar{\mathbf{yf}}^k = 0 \end{aligned} \tag{A.12}$$

where $\mathbf{yf}$ is a vector consisting of thirty elements $yf_i, i = 1, \ldots, 30$, and the solution of Eq. A.12 is denoted by $\hat{V}^k$, $\hat{R}^k$, and $\hat{\omega}\mathbf{f}^k$. When applying Algorithm I, the kth master problem Eq. 2.16 for this example is:

$$\begin{aligned} \min_{\bar{\mathbf{yf}},\eta} \quad & \eta \\ s.t. \quad & \eta \geq ISE(\hat{V},\hat{R},\hat{\omega}\mathbf{f}^k) + \boldsymbol{\zeta}^{k^T}(\hat{\omega}\mathbf{f}^k - \bar{\mathbf{yf}}) \end{aligned} \tag{A.13}$$

where the Lagrange multipliers $\boldsymbol{\zeta}^k$ satisfy the Karush–Kuhn–Tucker (KKT) optimality conditions of Eq. A.12:

$$\left.\frac{\partial(ISE)}{\partial V}\right|_k + \lambda_1^k \left.\frac{\partial g_1}{\partial V}\right|_k + \lambda_2^k \left.\frac{\partial g_2}{\partial V}\right|_k = 0 \tag{A.14}$$

$$\left.\frac{\partial(ISE)}{\partial R}\right|_k + \lambda_1^k \left.\frac{\partial g_1}{\partial R}\right|_k + \lambda_2^k \left.\frac{\partial g_2}{\partial R}\right|_k = 0 \tag{A.15}$$

$$\left.\frac{\partial(ISE)}{\partial(\mathbf{yf})}\right|_k + \lambda_1^k \left.\frac{\partial g_1}{\partial(\mathbf{yf})}\right|_k + \lambda_2^k \left.\frac{\partial g_2}{\partial(\mathbf{yf})}\right|_k = 0 \tag{A.16}$$

Now assume that at the solution of the primal problem Eq. A.12, inequality constraint $g_1 \leq 0$ is inactive but $g_2 \leq 0$ is active. In this case, $\lambda_1^k = 0$; eliminating λ_2^k from Eqs. A.14 and A.16 then gives:

$$\boldsymbol{\zeta}^k = -\left.\frac{\partial(ISE)}{\partial(\mathbf{yf})}\right|_k + \left.\frac{\frac{\partial(ISE)}{\partial V}\frac{\partial g_2}{\partial(\mathbf{yf})}}{\frac{\partial g_2}{\partial V}}\right|_k \tag{A.17}$$

Substituting Eq. A.17 in Eq. A.13 thus gives the following master problem for Algorithm I:

$$
\begin{aligned}
&\min_{\bar{\mathbf{yf}},\eta} \quad \eta \\
&s.t. \quad \eta \geq ISE(\hat{V}^k, \hat{R}^k, \hat{\boldsymbol{\omega}}\mathbf{f}^k) + \left.\frac{\partial(ISE)}{\partial(\mathbf{yf})}\right|_k (\bar{\mathbf{yf}} - \hat{\boldsymbol{\omega}}\mathbf{f}^k) \\
&\quad + \left.\frac{\frac{\partial(ISE)}{\partial V}\frac{\partial g_2}{\partial(\mathbf{yf})}}{\frac{\partial g_2}{\partial V}}\right|_k (\bar{\mathbf{yf}} - \hat{\boldsymbol{\omega}}\mathbf{f}^k) \\
&0 \geq g_1(\hat{V}^k, \hat{R}^k, \hat{\boldsymbol{\omega}}\mathbf{f}^k) + \left.\frac{\partial g_1}{\partial V}\right|_k (V - \hat{V}^k) + \left.\frac{\partial g_1}{\partial R}\right|_k (R - \hat{R}^k) + \left.\frac{\partial g_1}{\partial(\mathbf{yf})}\right|_k (\mathbf{yf} - \mathbf{yf}^k) \\
&0 \geq g_2(\hat{V}^k, \hat{R}^k, \hat{\boldsymbol{\omega}}\mathbf{f}^k) + \left.\frac{\partial g_2}{\partial V}\right|_k (V - \hat{V}^k) + \left.\frac{\partial g_2}{\partial R}\right|_k (R - \hat{R}^k) + \left.\frac{\partial g_2}{\partial(\mathbf{yf})}\right|_k (\mathbf{yf} - \mathbf{yf}^k)
\end{aligned}
\tag{A.18}
$$

When applying the OA-based, Algorithm II, the kth master problem Eq. 2.21 for this example is:

$$
\begin{aligned}
&\min_{\bar{\mathbf{yf}},V,R,\mathbf{yf},\eta} \quad \eta \\
&s.t. \quad \eta \geq ISE(\hat{V}^k, \hat{R}^k, \hat{\boldsymbol{\omega}}\mathbf{f}^k) + \left.\frac{\partial(ISE)}{\partial V}\right|_k (V - \hat{V}^k) + \left.\frac{\partial(ISE)}{\partial R}\right|_k (R - \hat{R}^k) \\
&\quad + \left.\frac{\partial(ISE)}{\partial(\mathbf{yf})}\right|_k (\mathbf{yf} - \hat{\boldsymbol{\omega}}\mathbf{f}^k) \\
&0 \geq g_1(\hat{V}^k, \hat{R}^k, \hat{\boldsymbol{\omega}}\mathbf{f}^k) + \left.\frac{\partial g_1}{\partial V}\right|_k (V - \hat{V}^k) + \left.\frac{\partial g_1}{\partial R}\right|_k (R - \hat{R}^k) + \left.\frac{\partial g_1}{\partial(\mathbf{yf})}\right|_k (\mathbf{yf} - \mathbf{yf}^k) \\
&0 \geq g_2(\hat{V}^k, \hat{R}^k, \hat{\boldsymbol{\omega}}\mathbf{f}^k) + \left.\frac{\partial g_2}{\partial V}\right|_k (V - \hat{V}^k) + \left.\frac{\partial g_2}{\partial R}\right|_k (R - \hat{R}^k) + \left.\frac{\partial g_2}{\partial(\mathbf{yf})}\right|_k (\mathbf{yf} - \mathbf{yf}^k) \\
&0 = \mathbf{yf} - \bar{\mathbf{yf}}
\end{aligned}
\tag{A.19}
$$

Now consider the case where, at the solution of the master problem above, the linearized constraint associated with g_2 is active but the constraint associated with g_1 remains inactive. In this case, the latter constraint can be ignored, while the equality of the former constraint implies that:

$$
V - \hat{V}^k = \frac{-\left.\frac{\partial g_2}{\partial(\mathbf{yf})}\right|_k (\mathbf{yf} - \hat{\boldsymbol{\omega}}\mathbf{f}^k) - \left.\frac{\partial g_2}{\partial R}\right|_k (R - \hat{R}^k)}{\left.\frac{\partial g_2}{V}\right|_k}
\tag{A.20}
$$

Utilizing Eq. A.20 in Eq. A.19 thus gives the following master problem for Algorithm II:

$$\begin{aligned}
&\min_{\bar{\mathbf{yf}}} \quad \eta \\
&s.t. \quad \eta \geq ISE(\hat{V}^k, \hat{R}^k, \hat{\omega}\mathbf{f}^k) + \left.\frac{\partial(ISE)}{\partial(\bar{\mathbf{yf}})}\right|_k (\bar{\mathbf{yf}} - \hat{\omega}\mathbf{f}^k) \\
&\quad + \left.\frac{\frac{\partial(ISE)}{\partial V}\frac{\partial g_2}{\partial(\bar{\mathbf{yf}})}}{\frac{\partial g_2}{\partial V}}\right|_k (\bar{\mathbf{yf}} - \hat{\omega}\mathbf{f}^k) \\
&\quad + \left[\left.\frac{\partial(ISE)}{\partial R}\right|_k - \left.\frac{\frac{\partial(ISE)}{\partial V}\frac{\partial g_2}{\partial R}}{\frac{\partial g_2}{\partial V}}\right|_k\right](R - \hat{R}^k)
\end{aligned} \tag{A.21}$$

However, the KKT optimality conditions Eqs. A.14 and A.15 with $\lambda_1^k = 0$ imply that:

$$\left.\frac{\partial(ISE)}{\partial R}\right|_k - \left.\frac{\frac{\partial(ISE)}{\partial V}\frac{\partial g_2}{\partial R}}{\frac{\partial g_2}{\partial V}}\right|_k = 0 \tag{A.22}$$

Thus, the master problem for Algorithm II becomes:

$$\begin{aligned}
&\min_{\bar{\mathbf{yf}}^k, \eta} \quad \eta \\
&s.t. \quad \eta \geq ISE(\hat{V}^k, \hat{R}^k, \hat{\omega}\mathbf{f}^k) + \left.\frac{\partial(ISE)}{\partial(\bar{\mathbf{yf}})}\right|_k (\bar{\mathbf{yf}} - \hat{\omega}\mathbf{f}^k) \\
&\quad + \left.\frac{\frac{\partial(ISE)}{\partial V}\frac{\partial g_2}{\partial(\bar{\mathbf{yf}})}}{\frac{\partial g_2}{\partial V}}\right|_k (\bar{\mathbf{yf}} - \hat{\omega}\mathbf{f}^k)
\end{aligned} \tag{A.23}$$

which is identical to the master problem for Algorithm I, Eq. A.18. Hence, if g_2 is active at the primal solution and is still the only active inequality constraint at the solution of the master problem, Algorithm II will give an identical solution to the master problem of Algorithm I. However, if any inequalities become active at the master solution that were inactive at the primal solution, then the master problem of Algorithm II will give a tighter lower bound than that of Algorithm I. This behaviour is indeed observed for the given example, as shown in Table A.1.

Table A.1 Comparison of the performance of the master problems of algorithms I and II for the example problem

Iteration number	1	2	3	4
Feed tray				
Algorithm I	26	24	25	23
Algorithm II	26	24	26	23
Lower bound, LB				
Algorithm I, LB_I	0.1576	0.1813	0.1815	0.1848
Algorithm II, LB_{II}	0.1576	0.1813	0.1815	0.2093
Active inequalities, $\mathbf{g}_j^k$				
Algorithm I	$\mathbf{g}_1^1$	$\mathbf{g}_1^2$	$\mathbf{g}_1^2$	$\mathbf{g}_1^3$
Algorithm II	$\mathbf{g}_1^1$	$\mathbf{g}_1^2$	$\mathbf{g}_1^2$	$\mathbf{g}_1^1, \mathbf{g}_1^2$
$LB_I \stackrel{?}{=} LB_{II}$	Yes	Yes	Yes	No

Supporting Information for Chap. 4

B

B.1 Nomenclature for the Example Models

See Tables B.1, B.2 and B.3.

B.2 Dynamic Model of the Reactive Distillation Column

Hereafter we present the full set of equations for the modeling of the reactive distillation column [151, 417, 486, 493].

B.2.1 Process Structure

The general column superstructure, similar to the binary distillation column example, enables the number of trays and feed tray location to be optimally determined via the following modeling formulation:

$$\sum_{k=1}^{N} yf_{feed,k} = 1 \tag{B.1}$$

$$\sum_{k=1}^{N} yr_k = 1 \tag{B.2}$$

B. Burnak et al., *Integrated Process Design and Operational Optimization via Multi-parametric Programming*, Synthesis Lectures on Engineering, Science, and Technology, https://doi.org/10.1007/978-3-032-10181-5

Table B.1 Nomenclature for the distillation column model

Symbol	Physical meaning
A_R	Area of reboiler
A_{tray}	Area of tray
B	Bottoms flow rate
C_{cond}	Cost of the condenser
C_{column}	Cost of the column
C_{reb}	Cost of the reboiler
D	Distillate flow rate
D_c	Diameter of distillation column
F	Volumetric inlet flow rate value
F_k	Volumetric inlet flow rate at the feed tray (k)
H_{weir}	Height of weir
L_k	Liquid flow rate at the kth tray
L_{weir}	Length of weir
$Level_k$	Liquid level at the kth tray
$M_{i,B}$	Molar hold up of ith species at the reboiler
$M_{i,D}$	Molar hold up of ith species at the condenser
$M_{i,k}$	Molar hold up of ith species at the kth tray
$M \& S$	Marshall and swift cost index
$Ntrays$	Maximum number of trays in the superstructure
$OpCost$	Operating cost of the column
P	Pressure of the column
$P^0_{benz,B}$	Saturation pressure of benzene in reboiler
$P^0_{benz,D}$	Saturation pressure of benzene in reflux drum
$P^0_{tol,B}$	Saturation pressure of toluene in reboiler
$P^0_{tol,D}$	Saturation pressure of toluene in reflux drum
R	Reflux flow rate value
R_k	Reflux flow rate at the reflux tray (k)
$TotalCost$	Total cost of the column
V_B	Vapor flow rate at the reboiler
V_D	Vapor flow rate at the condenser
V_k	Vapor flow rate at the kth tray
Vol_k	Liquid volume at the kth tray
$x_{i,B}$	Liquid mole fraction of species i in reboiler
$x_{i,D}$	Liquid mole fraction of species i in distillate
$x_{i,k}$	Liquid mole fraction of species i at the kth tray
$y_{i,B}$	Vapor mole fraction of species i in reboiler
$y_{i,D}$	Vapor mole fraction of species i in distillate
$y_{i,k}$	Vapor mole fraction of species i at the kth tray
$z_{i,f}$	Benzene composition at the feed tray
α	Constant relative volatility
$\rho_{Lmix,k}$	Density of liquid mixture at the kth tray
δ	Binary variables associated with the feed and reflux tray location

Table B.2 Nomenclature for the reactive distillation column model

Symbol	Physical meaning
a_i	Activity of component i
$A_{col,k}^{min}$	Minimum allowable column area for tray k
A_{col}	Cross-sectional area of the column
A_{holes}	Total area of all active holes
$A_{net,k}^{min}$	Minimum net area for vapor-liquid disengagement on tray k
A_{tray}	Active tray area
$Catalyst_k$	Catalyst mass for tray k
CH	Control horizon for model predictive controller
CR	Critical region
d	Disturbances to the process
D_{col}	Column diameter
$D_{col}^{min,k}$	Minimum allowable column diameter for tray k
De	Design variables
$Eff_{i,k}$	Murphree tray efficiency for component i on tray k
ent_k	Fractional entrainment for tray k
F	Objective function for dynamic optimization
F_{feed}	Feed flowrate
FLV_k	Sherwood flow parameter for tray k
$h^{l,v}$	Molar liquid or vapor enthalpy
$Height_{weir}$	Tray weir height
HS	Heightstack of the column
k	Reaction rate constant
$K1_k$	Empirical flooding velocity coefficient for tray k
K_a	Reaction equilibrium constant
L_k	Liquid molar flow rate for tray k
$Length_{weir}$	Tray weir length
$level_k$	Liquid level for tray k
$M_k^{l,v}$	Molar liquid or vapor holdup for tray k
$M_{i,k}$	Molar holdup of component i for tray k
OH	Output horizon of model predictive controller
P_k	Pressure for tray k
R	Reflux flow rate
r	Molar reaction rate
T	Temperature
u	Manipulated variable of the process
U_k	Molar internal energy holdup for tray k
V_b	Column boil up flow rate
V_k	Vapor molar flow rate for tray k
vel_k^{flood}	Flooding vapor velocity for tray k
vel_k	Vapor velocity for tray k
Vol_{tray}	Tray volume
x	Differential state
$x_{i,k}$	Liquid molar fraction of component i on tray k

(continued)

Table B.2 (continued)

Symbol	Physical meaning
Y	Binary decision variable
y	Controller output variable
$y_{i,k}$	Vapor molar fraction of component i for tray k
$y^*_{i,k}$	Equilibrium vapor molar fraction of component i
yf_k	
yr_k	
$z_{i,feed}$	
i	Index set for the components
k	Index set for the trays
ν	Stoichiometric coefficient
$\Phi^{l,v}$	Liquid or vapor fugacity coefficient
$\rho^{l,v}$	Liquid or vapor molar density
σ^l	Liquid surface tension
τ	Time horizon for dynamic optimization
θ	Parameter
$\bar{\rho}^{l,v}$	Liquid or vapor mass density

where $feed$ is the index set for feed streams, k is the index set for trays, N is the maximum number of column trays available which provides a reasonable estimate of the upper bound of number of trays, $yf_{feed,k}$ and yr_k are binary variables to denote if tray k is receiving (or not) feed or reflux. If $yf_{feed,k} = 1$ then all feed f enters tray k; similarly $yr_k = 1$ indicates that all reflux enters tray k (otherwise the binary variables take the value of 0). Also note that with Eqs. B.1 and B.2, no feed or reflux splitting is considered. The reflux is also constrained via Eq. B.3 to enter a tray below the feed:

$$\sum_{f} yf_{feed,k} - \sum_{k'=1}^{k} yr_{k'} \leq 0, \qquad k = 1, \ldots, N \tag{B.3}$$

B.2.2 Tray Modeling

The following assumptions are made in the current sieve tray model [151, 417]:

- Liquid and vapor phases are well-mixed;
- Liquid and vapor phases are in thermal and mechanical equilibrium with each other;
- Negligible downcomer dynamics;
- Negligible entrainment, weeping, draw-offs or external heat inputs for the trays.

The tray modeling equations are presented in what follows:

Table B.3 Nomenclature for the domestic CHP model

Symbol	Physical meaning
m_{th}	Mass of air entering the throttle valve
c_d	Valve discharge coefficient
A_{th}	Open area of the throttle valve
P_{ab}	Ambient environment pressure
R_β	Ideal gas constant
T_{ab}	Ambient environment temperature
$\psi\left(\frac{P_{ab}}{P_{mn}}\right)$	Laminar and turbulent flow indicator
m_{mn}	Mass of fluid in the manifolds
$m_{mn,in}$	Mass of fluid entering the manifolds
$m_{mn,out}$	Mass of fluid exiting the manifolds
$V_{mn,out}$	Volume of fluid exiting the manifolds
cpf	Pressure—flow coefficient
P_{mn}	Pressure in the manifolds
$P_{mn,out}$	Pressure at the outlet of the manifold
V_{mn}	Volume of fluid in the manifold
E	Internal energy
$h_{mn,in}$	Mass specific enthalpy of fluid entering the manifold
$h_{mn,out}$	Mass specific enthalpy of fluid exiting the manifold
m_{ex}	Mass of exhaust gasses
T_{mn}	Temperature in the manifold
V_d	ICE displacement volume
η_{vl}	Volumetric efficiency
ω_{en}	ICE angular velocity
$x_{air,i}$	Mass fraction of air components
$h^o_{f,air,i}$	Mass specific enthalpy of formation of air components
m_ϕ	Mass of fuel
h_ϕ	Enthalpy of fuel
$x_{\phi,j}$	Mass fraction of fuel components
$h^o_{f,fuel,j}$	Mass specific enthalpy of formation of fuel components
h_{ex}	Enthalpy of exhaust gases
$x_{ex,k}$	Mass fraction of exhaust gasses components
$h^o_{f,ex,k}$	Mass specific enthalpy of formation of exhaust components
Q_f	Heat of friction
$Q_{cg\to cw}$	Heat from cylinder gaskets to cylinder walls
W_c	Work of compression
W_{en}	Usable ICE work
To_{en}	Usable ICE torque
P_{meb}	Mean effective break pressure
To_{cl}	Flywheel torque
Fl	Flywheel inertia
Pec	Electric power
η_{en}	ICE efficiency
T_i	Temperature of system equipment *i i*

(continued)

Table B.3 (continued)

Symbol	Physical meaning
$Q_{in,i}$	Heat transferred into system equipment i
$Q_{out,i}$	Heat transferred out of system equipment
m_i	Mass of system equipment i
$c_{p,i}$	Mass specific heat capacity of system equipment i
$Q_{a\rightarrow b}$	Heat transfer from system component a to b
TC_{ab}	Heat transfer rate coefficient from ICE to ambient environment
A_{ab}	Area of heat transfer from ICE to ambient environment
T_a	Temperature of fluid a
T_b	Temperature of fluid b
UA	Overall heat transfer coefficient multiplied by the area of the heat exchangers
ΔT_{mean}	Logarithmic mean temperature
ΔT_{in}	Temperature difference between the inlet streams of the heat exchangers
ΔT_{out}	Temperature difference between the outlet streams of the heat exchangers

- Component molar balances:

$$\left(\sum_{k'=k}^{N} yr_{k'}\right)\cdot\frac{dM_{i,k}}{dt}=\sum_{feed}F_{feed,k}z_{i,feed}+R_kx_{i,d}+L_{k-1}x_{i,k-1}+V_{k+1}y_{i,k+1}$$
$$-L_kx_{i,k}+V_ky_{i,k}+\upsilon_iRate_k\quad i=1,\ldots NC,\ k=1,\ldots,N \tag{B.4}$$

where i is the index set for components, $M_{i,k}$ refers to the molar holdup of component i, $F_{feed,k}$ gives the flowrate of $feed$ to tray k, $z_{i,feed}$ is the molar fraction of component i in inlet stream $feed$, R_k is the reflux flow to tray k, L_k and V_k refer to liquid and vapor flowrates from tray k respectively, x_i and y_i are the molar fractions of component i in the liquid and vapor outlet streams from tray k, υ_i is the reaction stoichiometric coefficient, and $Rate_k$ gives the rate of reaction on tray k determined via specific reaction kinetics (e.g., Eq. 4.10):

$$Rate_k=r(P,T,\mathbf{x}) \tag{B.5}$$

- Energy balances (note that if constant column pressure is assumed, the energy balances are considered at steady-state):

$$\left(\sum_{k'=k}^{N} yr_{k'}\right)\cdot\frac{dU_k}{dt}=\sum_{feed}F_{feed,k}h_{feed}+R_kh_d^l+L_{k-1}h_{k-1}^l+V_{k+1}h_{k+1}^v \tag{B.6}$$
$$-L_kh_k^l+V_kh_k^v\qquad k=1,\ldots,N$$

where U_k denotes the internal energy holdup for tray k, h_k^l and h_k^v respectively refer to molar liquid and vapor molar enthalpies. Note that an additional term of heat of reaction is not needed if enthalpies are calculated on element-basis, while essential for component-based enthalpy calculation.

- Component molar holdups:

$$M_{i,k} = M_k^l x_{i,k} + M_k^v y_{i,k} \qquad i = 1, \ldots NC, \; k = 1, \ldots, N \tag{B.7}$$

where M_k^l (or M_k^l) is the total molar liquid (or vapor) holdup for tray k.

- Energy holdups:

$$U_k = M_k^l h_k^l + M_k^v h_k^v - 0.1 P_k Vol_{tray} \qquad k = 1, \ldots, N \tag{B.8}$$

where Vol_{tray} stands for tray volume, P_k is the stage pressure. In the MTBE reactive distillation study considered in this work, a constant column pressure profile is assumed.

- Volume constraints:

$$\frac{M_k^l}{\rho_k^l} + \frac{M_k^v}{\rho_k^v} = Vol_{tray} \qquad k = 1, \ldots, N \tag{B.9}$$

where ρ represents molar density.

- Equilibrium vapor phase composition:

$$\Phi_{i,k}^v y_{i,k}^* = \Phi_{i,k}^l x_{i,k} \qquad i = 1, \ldots NC, \; k = 1, \ldots, N \tag{B.10}$$

where Φ defines the vapor or liquid fugacity coefficient or activity coefficient.

- Murphree tray efficiency definition:

$$y_{i,k} = y_{i,k+1} + Eff_{i,k} \cdot (y_{i,k}^* - y_{i,k+1}) \qquad i = 1, \ldots NC, \; k = 1, \ldots, N \tag{B.11}$$

where $Eff_{i,k}$ stands for the Murphree tray efficiency.

- Molar fraction normalization:

$$\sum_{i=1}^{NC} x_{i,k} = \sum_{i=1}^{NC} y_{i,k} = 1 \qquad k = 1, \ldots, N \tag{B.12}$$

- Liquid levels:

$$Level_k = \frac{M_k^l}{\rho_k^l A_{tray}} \qquad k = 1, \ldots, N \tag{B.13}$$

where $Level_k$ gives the liquid level on tray k, A_{tray} denotes column tray area.

- Liquid outlet flowrates (modified Francis formula for liquid flow over a rectangular weir):

$$L_k = \begin{cases} 0, \text{ if } Level_k \le Height_{weir} \\ 1.84 \cdot \rho_k^l \cdot Length_{weir} \cdot (Level_k - Height_{weir})^{1.5}, \text{ otherwise} \end{cases} \tag{B.14}$$

- Pressure driving force for vapor inlet:

$$P_{k+1} - P_k = (\sum_{k'=k}^{N} yr_{k'}) \cdot (vel_{k+1}^2 \cdot \tilde{\rho}_{k+1}^v + \tilde{\rho}_k^l \cdot g \cdot Level_k) \tag{B.15}$$

where $\tilde{\rho}$ refer to mass density, g is the gravity constant, vel_k is the verlocity of vapor leaving tray k.

- Vapor velocity calculation:

$$vel_k = \frac{V_k}{\rho_k^v A_{holes}} \quad k = 1, \ldots, N \tag{B.16}$$

where A_{holes} refers to the total area of all active holes.

The following equations are used for tray geometry calculation:

- Free volume between trays:

$$Vol_{tray} = Space \cdot A_{tray} \tag{B.17}$$

- Cross-sectional area of the column

$$A_{col} = \frac{\pi}{4} D_{col}^2 \tag{B.18}$$

where $Space$ represents tray spacing, and D_{col} stands for column diameter. The other tray design parameters, such as $Length_{weir}$, $Height_{weir}$, $Active_{area}$, need to be specified.

The following equations are used for flooding and entrainment correlations:

- Fractional entrainment (80% flooding factor):

$$ent_k = 0.224exp(-2) + 2.377exp(-9.394FLV_k^{0.314}) \quad k = 1, \ldots, N \tag{B.19}$$

where ent_k is the fractional entrainment for tray k, FLV_k represents for Sherwood flow parameter for tray k.

- Sherwood flow parameter definition:

$$FLV_k = \frac{\tilde{L}_k}{\tilde{V}_k} \cdot (\frac{\tilde{\rho}_k^v}{\tilde{\rho}_k^l})^{0.5} \quad k = 1, \ldots, N \tag{B.20}$$

where the superscript˜denotes variables in mass basis.

- Mass flowrates:

$$\tilde{L}_k = L_k \cdot \sum_{i=1}^{NC} x_{i,k} MW_i \quad k = 1, \ldots, N \tag{B.21}$$

$$\tilde{V}_k = V_k \cdot \sum_{i=1}^{NC} y_{i,k} MW_i \quad k = 1, \ldots, N \tag{B.22}$$

- Flooding velocity:

$$vel_k^{flood} = (\frac{\sigma_k^l}{20})^{0.2} \cdot K1_k \cdot (\frac{\tilde{\rho}_k^l - \tilde{\rho}_k^v}{\tilde{\rho}_k^v})^{0.5} \quad k = 1, \ldots, N \tag{B.23}$$

where σ_k^l is surface liquid tension, $K1_k$ is an empirical coefficient given by:

$$K1_k = 0.0105 + 0.1496 \cdot Space^{0.755} \cdot exp(-1.463FLV_k^{0.842}) \quad k = 1, \ldots, N \tag{B.24}$$

- Minimum column diameter and area:

$$D_{col,k}^{min} = (\frac{4A_{col,k}^{min}}{\pi})^{0.5} \quad k = 1, \ldots, N \tag{B.25}$$

$$A_{net,k}^{min} = 0.9 \times A_{col,k}^{min} \quad k = 1, \ldots, N \tag{B.26}$$

- Minimum net area for vapor-liquid disengagement

$$A_{net,k}^{min} = \frac{V_k}{0.8 \cdot \rho_k^v \cdot vel_k^{flood}} \quad k = 1, \ldots, N \tag{B.27}$$

B.2.3 Reboiler & Condenser Modeling

The modeling of reboiler and condenser is in an analogous way to that of column trays, but with addition of heat input considerations in energy balances and without the pressure drop equation, flooding or entrainment correlations.

B.2.4 Physical Properties

The above-presented column model equations are independent of the selection of physical property models. Thus, the required physical properties can be generally described as:

$$h^l = h^l(P, T, \mathbf{x}) \tag{B.28}$$

$$h^v = h^v(P, T, \mathbf{y}) \tag{B.29}$$

$$\rho^l = \rho^l(P, T, \mathbf{x}) \tag{B.30}$$

$$\rho^v = \rho^v(P, T, \mathbf{y}) \tag{B.31}$$

$$\tilde{\rho}^l = \tilde{\rho}^l(P, T, \mathbf{x}) \tag{B.32}$$

$$\tilde{\rho}^v = \tilde{\rho}^v(P, T, \mathbf{y}) \tag{B.33}$$

$$\sigma^l = \sigma^l(P, T, \mathbf{x}) \tag{B.34}$$

$$\Phi_i^l = \Phi^l(P, T, \mathbf{x}) \qquad i = 1, \ldots, NC \tag{B.35}$$

$$\Phi_i^v = \Phi^v(P, T, \mathbf{y}) \qquad i = 1, \ldots, NC \tag{B.36}$$

B.2.5 Initial Conditions

In the case that the process is initially at steady-state, the initial conditions are:

$$\frac{dM_{i,\sqcup}}{dt} |_{t=0}= 0, \qquad i = 1, \ldots, NC, \qquad \sqcup = \{k = 1, \ldots, N\} \tag{B.37}$$

$$\frac{dU_{\sqcup}}{dt} |_{t=0}= 0, \qquad \sqcup = \{k = 1, \ldots, N\} \tag{B.38}$$

Supporting Information for Chap. 5

C

C.1 State Space Models of the Open Loop and Closed Loop Systems

We provide the state space matrices derived to approximate the open loop high-fidelity CSTR model as follows.

$$A \cdot 10^4 = \begin{bmatrix} 9670 & 8.00 & -23.0 & -915 & -23.1 & -20.0 & 5.12 \\ 21.0 & 9650 & -119 & 83.6 & -793 & 47.5 & -389 \\ 74.9 & -75.3 & 8610 & 214 & -1060 & 347 & 1360 \\ 656 & 172 & 127 & 9540 & 104 & 2270 & -309 \\ 187 & 179 & 219 & 3.88 & 9850 & 879 & -398 \\ -1440 & -151 & -193 & -741 & -138 & 5650 & 148 \\ 149 & 1130 & -36.5 & -31.8 & -82.0 & 57.9 & 6950 \end{bmatrix}$$

$$B \cdot 10^3 = \begin{bmatrix} 2.35 & -1.47 & -1.82 & 3.36 \\ 1.46 & -1.36 & 3.38 & -1.76 \\ 2.69 & -5.12 & 7.85 & -7.13 \\ -7.75 & 5.44 & 16.2 & 6.43 \\ -3.56 & 2.36 & 5.31 & 1.37 \\ 15.1 & -13.6 & -34.1 & -5.63 \\ -12.8 & -40.8 & -11.0 & -3.00 \end{bmatrix}$$

$$C \cdot 10^8 = \begin{bmatrix} -537 & 95.4 \\ 532 & -105 \\ -5070 & 944 \\ -2500 & 415 \\ -89.6 & 6.44 \\ 4860 & -822 \\ 5890 & -1070 \end{bmatrix}$$

$$D \cdot 10^2 = \begin{bmatrix} 2.73 & -748 & 303 & 202 & -816 & 613 & 49.4 \\ 582 & 497 & 90.8 & -31.6 & -30.6 & 1.61 & 4.02 \\ -705 & 730 & 134 & 28.2 & -43.3 & 5.46 & 4.80 \end{bmatrix}$$

Similarly, the closed loop state space models are determined as follows.

B. Burnak et al., *Integrated Process Design and Operational Optimization via Multi-parametric Programming*, Synthesis Lectures on Engineering, Science, and Technology, https://doi.org/10.1007/978-3-032-10181-5

Surrogate model 1

$$x_{t+1} = \begin{bmatrix} 0.004 & -0.001 & 0.002 \\ -0.031 & -0.010 & 0.045 \\ -0.118 & -0.026 & 0.118 \end{bmatrix} x_t + \begin{bmatrix} -7.2 \\ -4.7 \\ -3.1 \end{bmatrix} 10^{-4} Q_{total,t} \tag{C.1}$$

$$y_t = \begin{bmatrix} 0.340 & -0.037 & 0.066 \\ 0.072 & -0.040 & 0.031 \\ 0.048 & -0.042 & 0.041 \end{bmatrix} x_t$$

Surrogate model 2

$$x_{t+1} = \begin{bmatrix} 0.045 & 0.027 & -0.012 \\ 0.089 & -0.022 & -0.035 \\ 0.027 & 0.021 & -0.092 \end{bmatrix} x_t + \begin{bmatrix} 2.4 \cdot 10^{-4} & 0.130 \\ 9.0 \cdot 10^{-5} & -0.749 \\ 2.9 \cdot 10^{-6} & -0.716 \end{bmatrix} \begin{bmatrix} Q_{total,t} \\ C^{SP}_{P_2,t} \end{bmatrix}$$

$$y_t = \begin{bmatrix} 0.105 & -0.038 & -0.018 \\ 0.738 & -1.005 & -0.381 \\ 0 & 0 & 0 \end{bmatrix} x_t \tag{C.2}$$

Surrogate model 3

$$x_{t+1} = \begin{bmatrix} -0.011 & -0.012 & -0.016 \\ -0.067 & 0.112 & 0.117 \\ 0.134 & -0.148 & 0.220 \end{bmatrix} x_t + \begin{bmatrix} 2.5 \cdot 10^{-4} & 0.171 \\ 9.8 \cdot 10^{-5} & -0.620 \\ -5.5 \cdot 10^{-5} & 0.192 \end{bmatrix} \begin{bmatrix} Q_{total,t} \\ C^{SP}_{P_3,t} \end{bmatrix} \tag{C.3}$$

$$y_t = \begin{bmatrix} 0.014 & -0.008 & 0.004 \\ 0 & 0 & 0 \\ 0.516 & -1.081 & 0.477 \end{bmatrix} x_t$$

where x_t is the identified states, and y_t is the product concentrations ($C_{P_j,t}$).

C.2 A Snapshot of the Simultaneous Decisions

We present an example of the simultaneous decisions by the scheduler, surrogate model, and the controller at the end of the 1st hour of the operation depicted in Fig. 5.5. Following are the states and the remaining parameters of the system at $t = 1h$.

$$W = \begin{bmatrix} 1.26 \\ 0.24 \\ 0.65 \end{bmatrix}, \; C_P = \begin{bmatrix} 0.24 \\ 1.5 \times 10^{-4} \\ 0 \end{bmatrix}, \; a_{R,t_c=-1} = \begin{bmatrix} 1 \\ 0 \\ 0 \end{bmatrix}$$
$$DR_{t_s=0} = \begin{bmatrix} 5.81 \\ 6.06 \\ 5.89 \end{bmatrix}, \; DR_{t_s=1} = \begin{bmatrix} 5.79 \\ 6.09 \\ 5.89 \end{bmatrix}, \; DR_{t_s=2} = \begin{bmatrix} 5.82 \\ 6.06 \\ 5.97 \end{bmatrix} \tag{C.4}$$

Table C.1 Optimal decisions for the system parameters given in Eq. C.4

Decision variable	Affine expression
$F_{total,t_s=0}$	$= -16.7W_2 + DR_{t_s=0,2} + DR_{t_s=1,2}$
$F_{total,t_s=1}$	$= -16.7W_3 + DR_{t_s=0,3} + DR_{t_s=1,3} + DR_{t_s=2,3}$
$F_{total,t_s=2}$	$= DR_{t_s=2,2}$
Q_{total}	$= 500$
$C_{P_1}^{SP}$	$= 0$
$C_{P_2}^{SP}$	$= 0.91C_{P_1} - 0.01C_{P_2} + 0.02$
$C_{P_3}^{SP}$	$= 0$
a_{R_1}	$= 0$
a_{R_2}	$= 0.55$
a_{R_3}	$= 0$

Table C.2 Optimal decisions at $t = 105$ m

Decision variable	Affine expression
Q_{total}	$- F_{total,t_s=0}/C_{P_2} + 1.59 \times 10^2$
$C_{P_1}^{SP}$	$= 0$
$C_{P_2}^{SP}$	$= 0.12$
$C_{P_3}^{SP}$	$= 0$
a_{R_1}	$= 1 - a_{R_2}$
a_{R_2}	$= 0.06C_{P_1} + 0.64C_{P_2} - 0.01C_{P_3} - 4.0 \times 10^{-5}Q_{total}$
	$-0.08C_{P_1}^{SP} - 0.67C_{P_2}^{SP} + 0.98a_{R,t_c=-1} - 0.04$
a_{R_3}	$= 0$

Locating the system for the given parameters in Eq. C.4, the optimal decisions can be evaluated from the corresponding affine functions, as presented in Table C.1.

Due to the transition from C_{P_1} to C_{P_2}, the surrogate and control actions are mostly saturated. To highlight the affine expressions, a snapshot from the production period ($t =$ 105 m) is also presented in Table C.2. Note that the product concentrations at $t = 105$ m are $C_P = [0.003\ 0.146\ 0]^T$.

Supporting Information for Chap. 6

D

D.1 Complete MIDO Formulation of Integrated Design, Scheduling, and Control Problem for Example 1

The mathematical representation of the integrated design, scheduling, and control problem for Example 1 is given by Eq. D.1, in the form of an MIDO formulation.

$$\min_{u,s,des} \overbrace{a + bV^n}^{\text{Fixed cost}} + \overbrace{\int_0^\tau \sum_{i \in P} \alpha_i(t) W_i(t) dt}^{\text{Operating cost}} \tag{D.1a}$$

$$s.t. \quad \frac{dC_i}{dt} = \frac{Q_i C_i^f - Q_{total} C_i}{V} + \sum_{j \in J} s_{i,j} \mathcal{R}_j, \quad i \in (R \cup P) \tag{D.1b}$$

$$\mathcal{R}_j = k_j \prod_{i \in Rxn} C_i^{\nu_j}, \quad j \in J \tag{D.1c}$$

$$Q_{total} = \sum_{i \in R} Q_i \tag{D.1d}$$

$$\frac{dW_i}{dt} = \begin{cases} Q_{total} C_i - DR_i, & \text{if } Pur_i \geq 0.90 \\ -DR_i, & \text{if } Pur_i < 0.90 \end{cases}, i \in P \tag{D.1e}$$

$$Pur_i = \frac{C_i}{\sum_{i \in P} C_i} \tag{D.1f}$$

$$a_i = \frac{Q_i}{Q_{total}}, i \in R \tag{D.1g}$$

$$des := [V] \tag{D.1h}$$

B. Burnak et al., *Integrated Process Design and Operational Optimization via Multi-parametric Programming*, Synthesis Lectures on Engineering, Science, and Technology, https://doi.org/10.1007/978-3-032-10181-5

$$\begin{aligned} F_{i,t_s}(\theta_s) = \arg\min_{u_{t_s}} \quad & \sum_i \sum_{t_s} \alpha_{i,t_s} W_{i,t_s} \\ s.t. \quad & W_{i,t_s+1} = W_{i,t_s} + \Delta t_s F_{i,t_s} - \Delta t_s DR_{i,t_s} \\ & \underline{F}(V) y_{i,t_s} \le F_{i,t_s} \le \overline{F}(V) y_{i,t_s} \\ & \sum y_{i,t} = 1 \\ & \underline{W} \le F_{i,t_s} \le \overline{W} \\ & \theta_s = [V, W_{i,t_s=0}, DR_{i,t_s}] \\ & t_s \in \{0, \Delta t_s, \ldots, N_s \Delta t_s\} \end{aligned} \tag{D.1i}$$

$$\begin{aligned} s = \arg\min \quad & \sum_{t_{sm}} \| Q_{total,t_{sm}} - \tilde{Q}_{total,t_{sm}} \|^2_{R'} + \sum_{t_{sm}} \| \varepsilon'_{t_{sm}} \|^2_{P1'} \\ s.t. \quad & x_{t_{sm}+1} = A_{sm} x_{t_{sm}} + B_{sm} s_{t_{sm}} + C_{sm} des \\ & \tilde{Q}_{total,t_{sm}} = \frac{\sum_i F_{i,t_{sm}}}{C^*_{t_{sm}=0}} \\ & \theta_{sm} = [C^T_{t_{sm}=0}, F^T_{i,t_{sm}}, des^T]^T \\ & \underline{s} \le s_{t_{sm}} \le \overline{s} \\ & \underline{C} \le C_{t_{sm}} \le \overline{C} \\ & t_{sm} \in \{0, \Delta t_{sm}, \ldots, N_{sm} \Delta t_{sm}\} \end{aligned} \tag{D.1j}$$

$$s := [Q_{total,t_{sm}}(\theta_{sm}), C^{SP}_{i,t_{sm}}(\theta_{sm})] \tag{D.1k}$$

$$\begin{aligned} u = \arg\min_{a_{i,t_c}, \varepsilon_{t_c}, z_p} \quad & \sum_{t_c=1}^{N_c-1} \| C_{i,t_c} - C^{SP}_{i,t_c} \|^2_{QR} + \sum_{t_c=0}^{M_c-1} \| \Delta u_{t_c} \|^2_{R1} + \sum_{t=1}^{N_c} \| \varepsilon_{t_c} \|^2_{P1} \\ s.t. \quad & x_{t_c+1} = A_c x_{t_c} + B_c u_{t_c} + C_c [d^T_{t_c}, s^T_{t_c}, des^T]^T \\ & \hat{C}_{i,t_c} = D_c x_{t_c} + E_c u_{t_c} + F_c [d^T_{t_c}, s^T_{t_c}, des^T]^T \\ & C_{i,t_c} = \hat{C}_{i,t_c} + e \\ & e = C_{i,t_c=0} - \hat{C}_{i,t_c=0} \\ & \underline{u} z_p \le u_{t_c} \le \overline{u} z_p \\ & \sum_{p \in NP} z_p = 1 \\ & - C^*_{t_c} + Pur_{min} \sum_i C_{i,t_c=0} \le -\varepsilon_{t_c} + \mathcal{M}(1 - Y^*_{t_c}) \\ & \underline{x} \le x_{t_c} \le \overline{x}, \quad \underline{C_i} \le C_{i,t_c} \le \overline{C}_i \\ & \underline{u} \le u_{t_c} \le \overline{u}, \quad \underline{\Delta u} \le \Delta u_{t_c} \le \overline{\Delta u} \\ & \theta = [x^T_{t_c=0}, u^T_{t_c=-1}, d^T_{t_c=0}, s^T_{t_c}, des^T]^T \\ & \forall t_c \in \{0, 1, \ldots, N_c - 1\}, z_p \in \{0, 1\}, i \in P \end{aligned} \tag{D.1l}$$

Table D.1 Parameters of the high-fidelity CSTR model

Reaction rate constants	Value	Reactant concentration at the feed	Value	Unit cost	Value
k_1	0.1	$C^f_{R_1}$	1.0	α_1	1.0
k_2	0.9	$C^f_{R_2}$	0.8	α_2	1.5
k_3	1.5	$C^f_{R_3}$	1.0	α_3	1.8

$$u := [a_{i,t_c}(\theta_c), \varepsilon_{t_c}(\theta_c), z_p(\theta_c)] \tag{D.1m}$$

The objective function, given by Eq. D.1a, takes into account the annualized investment cost as a function of the reactor volume, and the operating cost as a function of the stored product in the inventory. Equations D.1b–D.1h are the dynamic high fidelity model of the multiproduct CSTR. More specifically, Eq. D.1b is the mass balance for the set of reactants (R) and products (P), Eq. D.1c is the rate expression for all reactions (J), Eq. D.1d is the volume balance at the exit of the reactor (mixture density is assumed to be constant), Eq. D.1e is the dynamics of the inventory levels in the storage tanks, Eq. D.1f is the purity of product $i \in P$, and Eq. D.1g defines the volumetric fraction of the reactants at the inlet. The The scheduling decisions are governed by Eqs. D.1i and D.1j to determine the operating region, and the closed loop control is regulated by Eq. D.1l. The parameters of the high-fidelity CSTR model are provided in Table D.1.

Observe that the scheduling and control decisions are postulated as lower level optimization problems, nested in an MIDO problem. Due to the implicit nature of the lower level optimization problems, Eq. D.1 is a challanging class of problem. Multiparametric programming allows for an offline map of solutions of the lower level decisions that can be implemented exactly in the upper level optimization problem.

D.2 Approximate Models

D.2.1 Example 1—CSTR

Here, we provide the approximate model that represents the open loop dynamics of the CSTR used in Example 1. The closed form of the state space model is given in Eqs. D.2a and D.2b, and the corresponding matrices are provided in Eqs. D.2c–D.2f.

$$x_{t_c+1} = Ax_{t_c} + B\begin{bmatrix} u_{1,t_c} \\ u_{2,t_c} \\ u_{3,t_c} \\ u_{4,t_c} \end{bmatrix} + C\begin{bmatrix} Q_{total,t_c} \\ V \end{bmatrix} \tag{D.2a}$$

$$\hat{C}_{i,t_c} = Dx_{t_c}, \quad i \in P \tag{D.2b}$$

$$A \cdot 10^4 = \begin{bmatrix} 9670 & 8.00 & -23.0 & -915 & -23.1 & -20.0 & 5.12 \\ 21.0 & 9650 & -119 & 83.6 & -793 & 47.5 & -389 \\ 74.9 & -75.3 & 8610 & 214 & -1060 & 347 & 1360 \\ 656 & 172 & 127 & 9540 & 104 & 2270 & -309 \\ 187 & 179 & 219 & 3.88 & 9850 & 879 & -398 \\ -1440 & -151 & -193 & -741 & -138 & 5650 & 148 \\ 149 & 1130 & -36.5 & -31.8 & -82.0 & 57.9 & 6950 \end{bmatrix} \tag{D.2c}$$

$$B \cdot 10^3 = \begin{bmatrix} 2.35 & -1.47 & -1.82 & 3.36 \\ 1.46 & -1.36 & 3.38 & -1.76 \\ 2.69 & -5.12 & 7.85 & -7.13 \\ -7.75 & 5.44 & 16.2 & 6.43 \\ -3.56 & 2.36 & 5.31 & 1.37 \\ 15.1 & -13.6 & -34.1 & -5.63 \\ -12.8 & -40.8 & -11.0 & -3.00 \end{bmatrix} \tag{D.2d}$$

$$C \cdot 10^8 = \begin{bmatrix} -537 & 95.4 \\ 532 & -105 \\ -5070 & 944 \\ -2500 & 415 \\ -89.6 & 6.44 \\ 4860 & -822 \\ 5890 & -1070 \end{bmatrix} \tag{D.2e}$$

$$D \cdot 10^2 = \begin{bmatrix} 2.73 & -748 & 303 & 202 & -816 & 613 & 49.4 \\ 582 & 497 & 90.8 & -31.6 & -30.6 & 1.61 & 4.02 \\ -705 & 730 & 134 & 28.2 & -43.3 & 5.46 & 4.80 \end{bmatrix} \tag{D.2f}$$

where x_{t_c} is the vector of identified states.

Similarly, the approximate models that represent the closed loop dynamics are given in Eqs. D.3–D.5. Note that the discretization time of the models are identical at $15min$.

$Surrogate\ model\ 1$

$$x_{t_{sm}+1} = \begin{bmatrix} 0.004 & -0.001 & 0.002 \\ -0.031 & -0.010 & 0.045 \\ -0.118 & -0.026 & 0.118 \end{bmatrix} x_{t_{sm}} + \begin{bmatrix} -7.2 \\ -4.7 \\ -3.1 \end{bmatrix} 10^{-4} Q_{total,t_{sm}} + \begin{bmatrix} 1.7 \\ 1.4 \\ 2.1 \end{bmatrix} 10^{-3} V$$

$$\hat{C}_{i,t_{sm}} = \begin{bmatrix} 0.340 & -0.037 & 0.066 \\ 0.072 & -0.040 & 0.031 \\ 0.048 & -0.042 & 0.041 \end{bmatrix} x_{t_{sm}}, \quad i \in P \tag{D.3}$$

Surrogate model 2

$$x_{t_{sm}+1} = \begin{bmatrix} 0.045 & 0.027 & -0.012 \\ 0.089 & -0.022 & -0.035 \\ 0.027 & 0.021 & -0.092 \end{bmatrix} x_{t_{sm}} + \begin{bmatrix} 2.4 \cdot 10^{-4} & 0.130 \\ 9.0 \cdot 10^{-5} & -0.749 \\ 2.9 \cdot 10^{-6} & -0.716 \end{bmatrix} \begin{bmatrix} Q_{total,t_{sm}} \\ C^{SP}_{P_2,t_{sm}} \end{bmatrix} + \begin{bmatrix} -3.8 \\ 0.2 \\ 3.5 \end{bmatrix} 10^{-5} V$$

$$\hat{C}_{i,t_{sm}} = \begin{bmatrix} 0.105 & -0.038 & -0.018 \\ 0.738 & -1.005 & -0.381 \\ 0 & 0 & 0 \end{bmatrix} x_{t_{sm}}, \quad i \in P \tag{D.4}$$

Surrogate model 3

$$x_{t_{sm}+1} = \begin{bmatrix} -0.011 & -0.012 & -0.016 \\ -0.067 & 0.112 & 0.117 \\ 0.134 & -0.148 & 0.220 \end{bmatrix} x_{t_{sm}} + \begin{bmatrix} 2.5 \cdot 10^{-4} & 0.171 \\ 9.8 \cdot 10^{-5} & -0.620 \\ -5.5 \cdot 10^{-5} & 0.192 \end{bmatrix} \begin{bmatrix} Q_{total,t_{sm}} \\ C^{SP}_{P_3,t_{sm}} \end{bmatrix} + \begin{bmatrix} -2.5 \\ 0.2 \\ -1.0 \end{bmatrix} 10^{-5} V$$

$$\hat{C}_{i,t_{sm}} = \begin{bmatrix} 0.014 & -0.008 & 0.004 \\ 0 & 0 & 0 \\ 0.516 & -1.081 & 0.477 \end{bmatrix} x_{t_{sm}}, \quad i \in P \tag{D.5}$$

The closed form of the approximate model used in the surrogate model formulation for the CHP system (Examples 3 and 4) is given in Eq. D.6.

$$\begin{bmatrix} E_{t_{sm}+1} \\ B_{t_{sm}+1} \end{bmatrix} = \begin{bmatrix} 1.0000 & 0 \\ 0.3880 & 0.9995 \end{bmatrix} \begin{bmatrix} E_{t_{sm}} \\ B_{t_{sm}} \end{bmatrix} + \begin{bmatrix} 0.9954 & 0 & 0 \\ -19.2613 & 0.1143 & -0.1143 \end{bmatrix} \begin{bmatrix} R_{t_{sm}} \\ Q_{t_{sm}} \\ D_{t_{sm}} \end{bmatrix} + \begin{bmatrix} 0 & 0 \\ -0.1143 & 0.0001 \end{bmatrix} \begin{bmatrix} \zeta^h_{t_{sm}} \\ V \end{bmatrix} \tag{D.6}$$

D.2.2 Example 5—Multipurpose batch process

The approximate state space model parameters for Reaction 1 is developed as follows.

$$A = \begin{bmatrix} 0.9905 & 0.0274 & -0.0299 \\ -0.1174 & 0.9713 & 0.0063 \\ -0.1309 & 0.0618 & 0.9084 \end{bmatrix}$$

$$B = \begin{bmatrix} -0.0860 & -0.1808 & -0.5156 \end{bmatrix}^T$$

$$D = \begin{bmatrix} 0.2735 & 0.2436 & -0.1495 \\ 1.2904 & 0.4882 & -0.4445 \end{bmatrix}$$

The approximate state space model for Reaction 2 has the following coefficient matrices.

$$A = \begin{bmatrix} 0.9797 & -0.0301 \\ -0.0684 & 0.8518 \end{bmatrix}$$

$$B = \begin{bmatrix} -0.0056 \\ -0.0024 \end{bmatrix}$$

$$D = \begin{bmatrix} -4.2018 & 5.0875 \\ -4.0276 & 0.3175 \end{bmatrix}$$

The approximate state space model for Reaction 3 has the following coefficient matrices.

$$A = \begin{bmatrix} 0.9863 & 0.0418 \\ 0.0195 & 0.9385 \end{bmatrix}$$

$$B = \begin{bmatrix} 0.0082 \\ 0.0019 \end{bmatrix}$$

$$D = \begin{bmatrix} 0.6002 & 0.7780 \\ 2.4079 & 1.6798 \end{bmatrix}$$

D.3 Cost Functions and Parameters for the Examples

For the CSTR cost functions used in Example 1 and 2, we use Eq. 6.3 to estimate the fixed design cost. The cost parameters a, b, and n are listed in Table D.2 for year 2010.

The cost estimation from 2010 is projected to 2018 by using Eq. D.7 the Chemical Engineering Plant Cost Index (CEPCI) [539].

$$Cost_{2018} = Cost_{2010} \frac{CEPCI_{2018}}{CEPCI_{2010}} \tag{D.7}$$

where the cost indexes $CEPCI_{2010}$ and $CEPCI_{2018}$ are 532.9 and 588.0, respectively.

Table D.2 Example 1—Reactor cost parameters for year 2010 [534]

Parameter	Value
a	61500
b	32500
n	0.6

For the CHP fixed cost estimation, on the other hand, we use a linear function, given in Eq. D.8 [85].

$$Cost_{CHP} = 370 + 0.0857 V_{CHP} \tag{D.8}$$

References

1. R. W. H. Sargent, "Integrated design and optimization of processes," *Chemical Engineering Progress*, vol. 63, no. 9, pp. 71–78, 1967.
2. B. Burnak, N. A. Diangelakis, and E. N. Pistikopoulos, "Towards the grand unification of process design, scheduling, and control—utopia or reality?," *Processes*, vol. 7, no. 7, 2019.
3. S. Terrazas-Moreno, A. Flores-Tlacuahuac, and I. E. Grossmann, "Simultaneous design, scheduling, and optimal control of a methyl-methacrylate continuous polymerization reactor," *AIChE Journal*, vol. 54, no. 12, pp. 3160–3170, 2008.
4. B. P. Patil, E. Maia, and L. A. Ricardez-sandoval, "Integration of Scheduling, Design, and Control of Multiproduct Chemical Processes Under Uncertainty," *AIChE Journal*, vol. 61, no. 8, 2015.
5. R. W. Koller and L. A. Ricardez-Sandoval, "A Dynamic Optimization Framework for Integration of Design, Control and Scheduling of Multi-product Chemical Processes under Disturbance and Uncertainty," *Computers & Chemical Engineering*, vol. 106, pp. 147–159, 2017.
6. R. W. Koller, L. A. Ricardez-Sandoval, and L. T. Biegler, "Stochastic back-off algorithm for simultaneous design, control, and scheduling of multiproduct systems under uncertainty," *AIChE Journal*, vol. 64, no. 7, pp. 2379–2389, 2018.
7. B. Burnak, N. A. Diangelakis, J. Katz, and E. N. Pistikopoulos, "Integrated process design, scheduling, and control using multiparametric programming," *Computers & Chemical Engineering*, vol. 125, pp. 164–184, 2019.
8. S. Qin and T. A. Badgwell, "A survey of industrial model predictive control technology," *Control Engineering Practice*, vol. 11, no. 7, pp. 733–764, 2003.
9. T. Takamatsu, I. Hashimoto, and H. Ohno, "Optimal design of a large complex system from the viewpoint of sensitivity analysis," *Industrial & Engineering Chemistry Process Design and Development*, vol. 9, no. 3, pp. 368–379, 1970.
10. N. Nishida, A. Ichikawa, and E. Tazaki, "Synthesis of optimal process systems with uncertainty," *Industrial & Engineering Chemistry Process Design and Development*, vol. 13, no. 3, pp. 209–214, 1974.
11. I. E. Grossmann and R. W. H. Sargent, "Optimum design of chemical plants with uncertain parameters," *AIChE Journal*, vol. 24, no. 6, pp. 1021–1028, 1978.
12. B. M. Kwak and E. J. Haug, "Optimum design in the presence of parametric uncertainty," *Journal of Optimization Theory and Applications*, vol. 19, pp. 527–546, 1976.

B. Burnak et al., *Integrated Process Design and Operational Optimization via Multi-parametric Programming*, Synthesis Lectures on Engineering, Science, and Technology, https://doi.org/10.1007/978-3-032-10181-5

13. K. P. Halemane and I. E. Grossmann, "Optimal process design under uncertainty," *AIChE Journal*, vol. 29, pp. 425–433, 5 1983.
14. R. E. Swaney and I. E. Grossmann, "An index for operational flexibility in chemical process design. part i: Formulation and theory," *AIChE Journal*, vol. 31, pp. 621–630, 4 1985.
15. R. E. Swaney and I. E. Grossmann, "An index for operational flexibility in chemical process design. part ii: Computational algorithms," *AIChE Journal*, vol. 31, pp. 631–641, 4 1985.
16. I. E. Grossmann and C. A. Floudas, "Active constraint strategy for flexibility analysis in chemical processes," *Computers & Chemical Engineering*, vol. 11, no. 6, pp. 675–693, 1987.
17. C. A. Floudas and I. E. Grossmann, "Synthesis of flexible heat exchanger networks with uncertain flowrates and temperatures," *Computers & Chemical Engineering*, vol. 11, no. 4, pp. 319–336, 1987.
18. Y. Shimizu, "A plain approach for dealing with flexibility problems in linear systems," *Computers & Chemical Engineering*, vol. 13, no. 10, pp. 1189–1191, 1989.
19. Y. Shimizu, "Application of flexibility analysis for compromise solution in large-scale linear systems," *Journal of Chemical Engineering of Japan*, vol. 22, no. 2, pp. 189–194, 1989.
20. V. Bansal, J. D. Perkins, and E. N. Pistikopoulos, "Flexibility analysis and design of linear systems by parametric programming," *AIChE Journal*, vol. 46, pp. 335–354, 2 2000.
21. E. N. Pistikopoulos and I. E. Grossmann, "Optimal retrofit design for improving process flexibility in linear systems," *Computers & Chemical Engineering*, vol. 12, no. 7, pp. 719–731, 1988. Special Issue on Process Systems Engineering.
22. E. N. Pistikopoulos and I. E. Grossmann, "Stochastic optimization of flexibility in retrofit design of linear systems," *Computers & Chemical Engineering*, vol. 12, no. 12, pp. 1215–1227, 1988.
23. E. N. Pistikopoulos and I. E. Grossmann, "Optimal retrofit design for improving process flexibility in nonlinear systems—i. fixed degree of flexibility," *Computers & Chemical Engineering*, vol. 13, no. 9, pp. 1003–1016, 1989.
24. E. N. Pistikopoulos and I. E. Grossmann, "Optimal retrofit design for improving process flexibility in nonlinear systems—ii. optimal level of flexibility," *Computers & Chemical Engineering*, vol. 13, no. 10, pp. 1087–1096, 1989.
25. E. N. Pistikopoulos and I. E. Grossmann, "Evaluation and redesign for improving flexibility in linear systems with infeasible nominal conditions," *Computers & Chemical Engineering*, vol. 12, no. 8, pp. 841–843, 1988.
26. C. Raspanti, J. Bandoni, and L. Biegler, "New strategies for flexibility analysis and design under uncertainty," *Computers & Chemical Engineering*, vol. 24, no. 9, pp. 2193–2209, 2000.
27. G. Kreisselmeier and R. Steinhauser, "Systematic control design by optimizing a vector performance index," in *Computer Aided Design of Control Systems* (M. CUENOD, ed.), pp. 113–117, Pergamon, 1980.
28. C. Chen and O. L. Mangasarian, "A class of smoothing functions for nonlinear and mixed complementarity problems," *Computational Optimization and Applications*, vol. 5, pp. 97–138, 1996.
29. E. N. Pistikopoulos and T. A. Mazzuchi, "A novel flexibility analysis approach for processes with stochastic parameters," *Computers & Chemical Engineering*, vol. 14, no. 9, pp. 991–1000, 1990.
30. D. A. Straub and I. E. Grossmann, "Integrated stochastic metric of flexibility for systems with discrete state and continuous parameter uncertainties," *Computers & Chemical Engineering*, vol. 14, no. 9, pp. 967–985, 1990.
31. D. A. Straub and I. E. Grossmann, "Design optimization of stochastic flexibility," *Computers & Chemical Engineering*, vol. 17, no. 4, pp. 339–354, 1993. An International Journal of Computer Applications in Chemical Engineering.
32. V. D. Dimitriadis and E. N. Pistikopoulos, "Flexibility analysis of dynamic systems," *Industrial & Engineering Chemistry Research*, vol. 34, no. 12, pp. 4451–4462, 1995.

33. H. Zhou, X. Li, Y. Qian, Y. Chen, and A. Kraslawski, "Optimizing the initial conditions to improve the dynamic flexibility of batch processes," *Industrial & Engineering Chemistry Research*, vol. 48, no. 13, pp. 6321–6326, 2009.
34. M. J. Mohideen, J. D. Perkins, and E. N. Pistikopoulos, "Optimal design of dynamic systems under uncertainty," *AIChE Journal*, vol. 42, pp. 2251–2272, 8 1996.
35. M. Mohideen, J. Perkins, and E. Pistikopoulos, "Optimal synthesis and design of dynamic systems under uncertainty," *Computers and Chemical Engineering*, vol. 20, no. SUPPL.2, pp. S895–S900, 1996.
36. M. Mohideen, J. Perkins, and E. Pistikopoulos, "Robust stability considerations in optimal design of dynamic systems under uncertainty," *Journal of Process Control*, vol. 7, no. 5, pp. 371–385, 1997.
37. A. D. Pretoro, L. Montastruc, F. Manenti, and X. Joulia, "Flexibility analysis of a distillation column: Indexes comparison and economic assessment," *Computers & Chemical Engineering*, vol. 124, pp. 93–108, 2019.
38. Y. Zhu, S. Legg, and C. D. Laird, "Optimal design of cryogenic air separation columns under uncertainty," *Computers & Chemical Engineering*, vol. 34, no. 9, pp. 1377–1384, 2010. Selected papers from the 7th International Conference on the Foundations of Computer-Aided Process Design (FOCAPD, 2009, Breckenridge, Colorado, USA.
39. W. Huang, X. Li, S. Yang, and Y. Qian, "Dynamic flexibility analysis of chemical reaction systems with time delay: Using a modified finite element collocation method," *Chemical Engineering Research and Design*, vol. 89, no. 10, pp. 1938–1946, 2011.
40. A. E. S. Konukman, M. C. Çamurdan, and U. Akman, "Simultaneous flexibility targeting and synthesis of minimum-utility heat-exchanger networks with superstructure-based milp formulation," *Chemical Engineering and Processing: Process Intensification*, vol. 41, no. 6, pp. 501–518, 2002.
41. A. E. S. Konukman and U. Akman, "Flexibility and operability analysis of a hen-integrated natural gas expander plant," *Chemical Engineering Science*, vol. 60, no. 24, pp. 7057–7074, 2005.
42. M. Escobar, J. O. Trierweiler, and I. E. Grossmann, "Simultaneous synthesis of heat exchanger networks with operability considerations: Flexibility and controllability," *Computers & Chemical Engineering*, vol. 55, pp. 158–180, 2013.
43. D. K. Varvarezos, I. E. Grossmann, and L. T. Biegler, "An outer-approximation method for multiperiod design optimization," *Industrial & Engineering Chemistry Research*, vol. 31, no. 6, pp. 1466–1477, 1992.
44. E. Pistikopoulos and M. Ierapetritou, "Novel approach for optimal process design under uncertainty," *Computers & Chemical Engineering*, vol. 19, no. 10, pp. 1089–1110, 1995.
45. M. Morari, "Design of resilient processing plants—iii: A general framework for the assessment of dynamic resilience," *Chemical Engineering Science*, vol. 38, no. 11, pp. 1881–1891, 1983.
46. M. Morari, "Flexibility and resiliency of process systems," *Computers & Chemical Engineering*, vol. 7, no. 4, pp. 423–437, 1983.
47. I. E. Grossmann and M. Morari, "Operability, Resiliency, and Flexibility: process design objectives for a changing world," 1 1983.
48. M. Morari, W. Grimm, M. J. Oglesby, and I. D. Prosser, "Design of resilient processing plants—vii. design of energy management system for unstable reactors—new insights," *Chemical Engineering Science*, vol. 40, no. 2, pp. 187–198, 1985.
49. A. Palazoglu, B. Manousiouthakis, and Y. Arkun, "Design of chemical plants with improved dynamic operability in an environment of uncertainty," *Industrial & Engineering Chemistry Process Design and Development*, vol. 24, no. 3, pp. 802–813, 1985.

50. A. Palazoglu and Y. Arkun, "A multiobjective approach to design chemical plants with robust dynamic operability characteristics," *Computers & Chemical Engineering*, vol. 10, no. 6, pp. 567–575, 1986.
51. S. Skogestad and M. Morari, "Design of resilient processing plants-ix. effect of model uncertainty on dynamic resilience," *Chemical Engineering Science*, vol. 42, no. 7, pp. 1765–1780, 1987.
52. R. D. Colberg, M. Morari, and D. W. Townsend, "A resilience target for heat exchanger network synthesis," *Computers & Chemical Engineering*, vol. 13, no. 7, pp. 821–837, 1989.
53. J. D. Perkins and M. P. F. Wong, "Assessing controllability of chemical plants," *Chemical Engineering Research and Design*, vol. 63, no. 6, pp. 358–362, 1985.
54. H. H. Rosenbrock, *State-space and multivariable theory*. Studies in dynamical systems series, Wiley Interscience Division, 1970.
55. P. Psarris and C. A. Floudas, "Improving dynamic operability in mimo systems with time delays," *Chemical Engineering Science*, vol. 45, no. 12, pp. 3505–3524, 1990.
56. P. Psarris and C. A. Floudas, "Dynamic operability of mimo systems with time delays and transmission zeroes—i. assessment," *Chemical Engineering Science*, vol. 46, no. 10, pp. 2691–2707, 1991.
57. P. Psarris and C. A. Floudas, "Dynamic operability of mimo systems with time delays and transmission zeroes—ii. enhancement," *Chemical Engineering Science*, vol. 46, no. 10, pp. 2709–2728, 1991.
58. G. W. Barton, W. K. Chan, and J. D. Perkins, "Interaction between process design and process control: the role of open-loop indicators," *Journal of Process Control*, vol. 1, no. 3, pp. 161–170, 1991.
59. L. Narraway, J. Perkins, and G. Barton, "Interaction between process design and process control: economic analysis of process dynamics," *Journal of Process Control*, vol. 1, no. 5, pp. 243–250, 1991.
60. L. Narraway and J. Perkins, "Selection of process control structure based on economics," *Computers & Chemical Engineering*, vol. 18, pp. S511–S515, 1994. European Symposium on Computer Aided Process Engineering—3.
61. P. Bahri, J. Bandoni, G. Barton, and J. Romagnoli, "Back-off calculations in optimising control: A dynamic approach," *Computers & Chemical Engineering*, vol. 19, pp. 699–708, 1995.
62. S. Walsh and J. Perkins, "Integrated design of effluent treatment systems," *IFAC Proceedings Volumes*, vol. 25, no. 24, pp. 107–112, 1992. IFAC Workshop on Interactions between Process Design and Process Control, London, UK, 7-8 September.
63. M. L. Luyben and C. A. Floudas, "A multiobjective optimization approach for analyzing the interaction of design and control," *IFAC Proceedings Volumes*, vol. 25, no. 24, pp. 101–106, 1992. IFAC Workshop on Interactions between Process Design and Process Control, London, UK, 7-8 September.
64. N. Shah, C. C. Pantelides, and R. W. H. Sargent, "The design and scheduling of multipurpose batch plants," *IFAC Proceedings Volumes*, vol. 25, no. 24, pp. 203–208, 1992. IFAC Workshop on Interactions between Process Design and Process Control, London, UK, 7-8 September.
65. E. Kondili, C. Pantelides, and R. Sargent, "A general algorithm for short-term scheduling of batch operations—i. milp formulation," *Computers & Chemical Engineering*, vol. 17, no. 2, pp. 211–227, 1993. An International Journal of Computer Applications in Chemical Enginnering.
66. T. V. Thomaidis and E. N. Pistikopoulos, "Design of flexible and reliable process systems," *IFAC Proceedings Volumes*, vol. 25, no. 24, pp. 235–240, 1992. IFAC Workshop on Interactions between Process Design and Process Control, London, UK, 7-8 September.

67. S. Walsh and J. Perkins, "Application of integrated process and control system design to waste water neutralisation," *Computers & Chemical Engineering*, vol. 18, pp. S183–S187, 1994. European Symposium on Computer Aided Process Engineering—3.
68. L. T. Narraway and J. D. Perkins, "Selection of process control structure based on linear dynamic economics," *Industrial & Engineering Chemistry Research*, vol. 32, no. 11, pp. 2681–2692, 1993.
69. M. L. Luyben and C. A. Floudas, "Analyzing the interaction of design and control—1. a multiobjective framework and application to binary distillation synthesis," *Computers & Chemical Engineering*, vol. 18, no. 10, pp. 933–969, 1994. An International Journal of Computer Application in Chemical Engineering.
70. M. L. Luyben and C. A. Fluodas, "Analyzing the interaction of design and control—2. reactor-separator-recycle system," *Computers & Chemical Engineering*, vol. 18, no. 10, pp. 971–993, 1994. An International Journal of Computer Application in Chemical Engineering.
71. V. Bansal, J. Perkins, E. Pistikopoulos, R. Ross, and J. Van Schijndel, "Simultaneous design and control optimisation under uncertainty," *Computers and Chemical Engineering*, vol. 24, no. 2-7, pp. 261–266, 2000.
72. L. Ricardez-Sandoval, H. Budman, and P. Douglas, "Simultaneous design and control of processes under uncertainty: A robust modelling approach," *Journal of Process Control*, vol. 18, no. 7-8, pp. 735–752, 2008.
73. L. A. Ricardez-Sandoval, H. M. Budman, and P. L. Douglas, "Application of robust control tools to the simultaneous design and control of dynamic systems," *Industrial & Engineering Chemistry Research*, vol. 48, no. 2, pp. 801–813, 2009.
74. L. A. Ricardez-Sandoval, H. M. Budman, and P. L. Douglas, "Simultaneous design and control of chemical processes with application to the tennessee eastman process," *Journal of Process Control*, vol. 19, no. 8, pp. 1377–1391, 2009. Special Section on Hybrid Systems: Modeling, Simulation and Optimization.
75. I. Kookos and J. Perkins, "Control structure selection based on economics: Generalization of the back-off methodology," *AIChE Journal*, vol. 62, no. 9, pp. 3056–3064, 2016.
76. S. Mehta and L. A. Ricardez-sandoval, "Integration of Design and Control of Dynamic Systems under Uncertainty : A New Back-O ff Approach," 2016.
77. M. Rafiei-Shishavan, S. Mehta, and L. Ricardez-Sandoval, "Simultaneous design and control under uncertainty: A back-off approach using power series expansions," *Computers and Chemical Engineering*, vol. 99, pp. 66–81, 2017.
78. M. Rafiei and L. A. Ricardez-Sandoval, "Stochastic back-off approach for integration of design and control under uncertainty," *Industrial & Engineering Chemistry Research*, vol. 57, no. 12, pp. 4351–4365, 2018.
79. I. K. Kookos and J. D. Perkins, "An algorithm for simultaneous process design and control," *Industrial & Engineering Chemistry Research*, vol. 40, no. 19, pp. 4079–4088, 2001.
80. A. Malcolm, J. Polan, L. Zhang, B. Ogunnaike, and A. Linninger, "Integrating systems design and control using dynamic flexibility analysis," *AIChE Journal*, vol. 53, no. 8, pp. 2048–2061, 2007.
81. J. Moon, S. Kim, and A. A. Linninger, "Integrated design and control under uncertainty: Embedded control optimization for plantwide processes," *Computers & Chemical Engineering*, vol. 35, no. 9, pp. 1718–1724, 2011. Energy Systems Engineering.
82. D. Brengel and W. Seider, "Coordinated design and control optimization of nonlinear processes," *Computers and Chemical Engineering*, vol. 16, no. 9, pp. 861–886, 1992.
83. A. Bemporad, M. Morari, V. Dua, and E. N. Pistikopoulos, "The explicit linear quadratic regulator for constrained systems," *Automatica*, vol. 38, no. 1, pp. 3–20, 2002.

84. V. Sakizlis, J. Perkins, and E. Pistikopoulos, "Parametric controllers in simultaneous process and control design optimization," *Industrial and Engineering Chemistry Research*, vol. 42, no. 20, pp. 4545–4563, 2003.
85. N. A. Diangelakis, B. Burnak, J. Katz, and E. N. Pistikopoulos, "Process design and control optimization: A simultaneous approach by multi-parametric programming," *AIChE Journal*, vol. 63, no. 11, pp. 4827–4846, 2017.
86. N. A. Diangelakis and E. N. Pistikopoulos, "A multi-scale energy systems engineering approach to residential combined heat and power systems," *Computers & Chemical Engineering*, vol. 102, pp. 128–138, 2017.
87. K. Sanchez-Sanchez and L. Ricardez-Sandoval, "Simultaneous design and control under uncertainty using model predictive control," *Industrial and Engineering Chemistry Research*, vol. 52, no. 13, pp. 4815–4833, 2013.
88. S. Engell and I. Harjunkoski, "Optimal operation: Scheduling, advanced control and their integration," *Computers and Chemical Engineering*, vol. 47, pp. 121–133, 2012.
89. A. Prata, J. Oldenburg, A. Kroll, and W. Marquardt, "Integrated scheduling and dynamic optimization of grade transitions for a continuous polymerization reactor," *Computers and Chemical Engineering*, vol. 32, no. 3, pp. 463–476, 2008.
90. A. Flores-Tlacuahuac and I. E. Grossmann, "Simultaneous cyclic scheduling and control of a multiproduct CSTR," *Industrial and Engineering Chemistry Research*, vol. 45, no. 20, pp. 6698–6712, 2006.
91. Y. Nie, L. T. Biegler, and J. M. Wassick, "Integrated Scheduling and Dynamic Optimization of Batch Processes Using State Equipment Networks," *AIChE Journal*, vol. 58, no. 11, pp. 3416–3432, 2012.
92. Y. Chu and F. You, "Integration of production scheduling and dynamic optimization for multi-product CSTRs: Generalized Benders decomposition coupled with global mixed-integer fractional programming," *Computers and Chemical Engineering*, vol. 58, pp. 315–333, 2013.
93. Y. Chu and F. You, "Integrated planning, scheduling, and dynamic optimization for batch processes: Minlp model formulation and efficient solution methods via surrogate modeling," *Industrial & Engineering Chemistry Research*, vol. 53, no. 34, pp. 13391–13411, 2014.
94. O. Palma-Flores and L. A. Ricardez-Sandoval, "Integration of design and nmpc-based control for chemical processes under uncertainty: An mpcc-based framework," *Computers & Chemical Engineering*, vol. 162, p. 107815, 2022.
95. B. Burnak, J. Katz, N. A. Diangelakis, and E. N. Pistikopoulos, "Simultaneous process scheduling and control: A multiparametric programming-based approach," *Industrial & Engineering Chemistry Research*, vol. 57, no. 11, pp. 3963–3976, 2018.
96. C. Tsay and M. Baldea, "Integrating production scheduling and process control using latent variable dynamic models," *Control Engineering Practice*, vol. 94, p. 104201, 2020.
97. J. Zhuge and M. G. Ierapetritou, "An integrated framework for scheduling and control using fast model predictive control," *AIChE Journal*, vol. 61, no. 10, pp. 3304–3319, 2015.
98. V. M. Charitopoulos, L. G. Papageorgiou, and V. Dua, "Closed-loop integration of planning, scheduling and multi-parametric nonlinear control," *Computers & Chemical Engineering*, vol. 122, pp. 172–192, 2019. 2017 Edition of the European Symposium on Computer Aided Process Engineering (ESCAPE-27).
99. M. Baldea, J. Du, J. Park, and I. Harjunkoski, "Integrated Production Scheduling and Model Predictive Control of Continuous Processes," *AIChE Journal*, vol. 61, no. 12, pp. 4179–4190, 2015.
100. J. M. Simkoff and M. Baldea, "Production scheduling and linear mpc: Complete integration via complementarity conditions," *Computers & Chemical Engineering*, vol. 125, pp. 287–305, 2019.

101. J. E. Remigio and C. L. Swartz, "Production scheduling in dynamic real-time optimization with closed-loop prediction," *Journal of Process Control*, vol. 89, pp. 95–107, 2020.
102. O. Andrés-Martínez and L. A. Ricardez-Sandoval, "Integration of planning, scheduling, and control: A review and new perspectives," *The Canadian Journal of Chemical Engineering*, vol. 100, no. 9, pp. 2057–2070, 2022.
103. M. Heidarinejad, J. Liu, and P. D. Christofides, "Economic model predictive control of nonlinear process systems using lyapunov techniques," *AIChE Journal*, vol. 58, no. 3, pp. 855–870, 2012.
104. M. Ellis, H. Durand, and P. D. Christofides, "A tutorial review of economic model predictive control methods," *Journal of Process Control*, vol. 24, no. 8, pp. 1156–1178, 2014. Economic nonlinear model predictive control.
105. M. Ellis and P. D. Christofides, "Optimal time-varying operation of nonlinear process systems with economic model predictive control," *Industrial & Engineering Chemistry Research*, vol. 53, no. 13, pp. 4991–5001, 2014.
106. K. Subramanian, J. B. Rawlings, and C. T. Maravelias, "Economic model predictive control for inventory management in supply chains," *Computers & Chemical Engineering*, vol. 64, pp. 71–80, 2014.
107. O. Santander, A. Elkamel, and H. Budman, "Economic model predictive control of chemical processes with parameter uncertainty," *Computers & Chemical Engineering*, vol. 95, pp. 10–20, 2016.
108. J. Zhuge and M. G. Ierapetritou, "Integration of scheduling and control for batch processes using multi-parametric model predictive control," *AIChE Journal*, vol. 60, no. 9, pp. 3169–3183, 2014.
109. Y. Nie, L. T. Biegler, C. M. Villa, and J. M. Wassick, "Discrete time formulation for the integration of scheduling and dynamic optimization," *Industrial & Engineering Chemistry Research*, vol. 54, no. 16, pp. 4303–4315, 2015.
110. C. T. Maravelias and I. E. Grossmann, "A hybrid milp/cp decomposition approach for the continuous time scheduling of multipurpose batch plants," *Computers & chemical engineering*, vol. 28, no. 10, pp. 1921–1949, 2004.
111. B. W. Overturf, G. V. Reklaitis, and J. M. Woods, "Gasp iv and the simulation of batch/semicontinuous operations: single train process," *Industrial & Engineering Chemistry Process Design and Development*, vol. 17, no. 2, pp. 161–165, 1978.
112. E. N. Pistikopoulos, S. S. Akundi, D. Kenefake, and N. A. Diangelakis, "The quest towards the integration of process control, process operations, and process operability –industrial need or academic curiosity?," *Computers & Chemical Engineering*, vol. 180, p. 108470, 2024.
113. R. Bindlish, "Scheduling, optimization and control of power for industrial cogeneration plants," *Computers & Chemical Engineering*, vol. 114, pp. 221–224, 2018.
114. D. Kenefake, I. Pappas, S. Avraamidou, B. Beykal, H. S. Ganesh, Y. Cao, Y. Wang, J. Otashu, S. Leyland, J. Flores-Cerrillo, *et al.*, "A smart manufacturing strategy for multiparametric model predictive control in air separation systems," *Journal of Advanced Manufacturing and Processing*, vol. 4, no. 4, p. e10120, 2022.
115. L. S. Dias, R. C. Pattison, C. Tsay, M. Baldea, and M. G. Ierapetritou, "A simulation-based optimization framework for integrating scheduling and model predictive control, and its application to air separation units," *Computers & Chemical Engineering*, vol. 113, pp. 139–151, 2018.
116. M. T. Kelley, C. Tsay, Y. Cao, Y. Wang, J. Flores-Cerrillo, and M. Baldea, "A data-driven linear formulation of the optimal demand response scheduling problem for an industrial air separation unit," *Chemical Engineering Science*, vol. 252, p. 117468, 2022.
117. G. Guillén-Gosálbez, F. D. Mele, and I. E. Grossmann, "A bi-criterion optimization approach for the design and planning of hydrogen supply chains for vehicle use," *AIChE Journal*, vol. 56, no. 3, pp. 650–667, 2010.

118. Y. Chu and F. You, "Integrated scheduling and control of a polymerization reactor with online closed-loop implementation," in *Computer Aided Chemical Engineering*, vol. 32, pp. 481–486, Elsevier, 2013.
119. D. Elixmann, J. Busch, and W. Marquardt, "Integration of model-predictive scheduling, dynamic real-time optimization and output tracking for a wastewater treatment process," *IFAC Proceedings Volumes*, vol. 43, no. 6, pp. 90–95, 2010.
120. P. Pravin, S. Misra, S. Bhartiya, and R. D. Gudi, "A reactive scheduling and control framework for integration of renewable energy sources with a reformer-based fuel cell system and an energy storage device," *Journal of Process Control*, vol. 87, pp. 147–165, 2020.
121. M. H. Sahraei and L. Ricardez-Sandoval, "Controllability and optimal scheduling of a co2 capture plant using model predictive control," *International Journal of Greenhouse Gas Control*, vol. 30, pp. 58–71, 2014.
122. P. Mathur, C. L. Swartz, D. Zyngier, and F. Welt, "Robust online scheduling for optimal short-term operation of cascaded hydropower systems under uncertainty," *Journal of Process Control*, vol. 98, pp. 52–65, 2021.
123. P. Mathur, C. L. Swartz, D. Zyngier, and F. Welt, "Uncertainty management via online scheduling for optimal short-term operation of cascaded hydropower systems," *Computers & Chemical Engineering*, vol. 134, p. 106677, 2020.
124. M. J. Risbeck, C. T. Maravelias, J. B. Rawlings, and R. D. Turney, "Mixed-integer optimization methods for online scheduling in large-scale hvac systems," *Optimization Letters*, vol. 14, pp. 889–924, 2020.
125. K. Subramanian, J. B. Rawlings, C. T. Maravelias, J. Flores-Cerrillo, and L. Megan, "Integration of control theory and scheduling methods for supply chain management," *Computers and Chemical Engineering*, vol. 51, pp. 4–20, 2013.
126. L. T. Biegler, "Advanced optimization strategies for integrated dynamic process operations," *Computers & Chemical Engineering*, vol. 114, pp. 3–13, 2018.
127. C. R. Touretzky and M. Baldea, "Integrating scheduling and control for economic mpc of buildings with energy storage," *Journal of Process Control*, vol. 24, no. 8, pp. 1292–1300, 2014. Economic nonlinear model predictive control.
128. K. Subramanian, C. T. Maravelias, and J. B. Rawlings, "A state-space model for chemical production scheduling," *Computers & Chemical Engineering*, vol. 47, pp. 97–110, 2012.
129. R. D. McAllister, J. B. Rawlings, and C. T. Maravelias, "Rescheduling penalties for economic model predictive control and closed-loop scheduling," *Industrial & Engineering Chemistry Research*, vol. 59, no. 6, pp. 2214–2228, 2019.
130. M. J. Risbeck, C. T. Maravelias, and J. B. Rawlings, "Unification of closed-loop scheduling and control: State-space formulations, terminal constraints, and nominal theoretical properties," *Computers & Chemical Engineering*, vol. 129, p. 106496, 2019.
131. M. Ierapetritou and E. Pistikopoulos, "Global optimization for stochastic planning, scheduling and design problems," *Global optimization in engineering design*, pp. 231–287, 1996.
132. M. A. Gutiérrez-Limón, A. Flores-Tlacuahuac, and I. E. Grossmann, "A Multiobjective Optimization Approach for the Simultaneous Single Line Scheduling and Control of CSTRs," *Industrial & Engineering Chemistry Research*, vol. 51, no. 17, pp. 5881–5890, 2012.
133. F. Albalawi, H. Durand, and P. D. Christofides, "Process operational safety using model predictive control based on a process safeness index," *Computers & Chemical Engineering*, vol. 104, pp. 76–88, 2017.
134. Z. Wu, F. Albalawi, Z. Zhang, J. Zhang, H. Durand, and P. D. Christofides, "Control lyapunov-barrier function-based model predictive control of nonlinear systems," *Automatica*, vol. 109, p. 108508, 2019.

135. F. Albalawi, H. Durand, and P. D. Christofides, "Process operational safety via model predictive control: Recent results and future research directions," *Computers & Chemical Engineering*, vol. 114, pp. 171–190, 2018.
136. P. Mhaskar, J. Liu, and P. D. Christofides, *Fault-tolerant process control: methods and applications*. Springer Science & Business Media, 2012.
137. P. Mhaskar, "Robust model predictive control design for fault-tolerant control of process systems," *Industrial & engineering chemistry research*, vol. 45, no. 25, pp. 8565–8574, 2006.
138. L. Lao, M. Ellis, and P. D. Christofides, "Proactive fault-tolerant model predictive control," *AIChE Journal*, vol. 59, no. 8, pp. 2810–2820, 2013.
139. M. Kettunen, P. Zhang, and S.-L. Jämsä-Jounela, "An embedded fault detection, isolation and accommodation system in a model predictive controller for an industrial benchmark process," *Computers & Chemical Engineering*, vol. 32, no. 12, pp. 2966–2985, 2008.
140. J. Prakash, S. C. Patwardhan, and S. Narasimhan, "A supervisory approach to fault-tolerant control of linear multivariable systems," *Industrial & Engineering Chemistry Research*, vol. 41, no. 9, pp. 2270–2281, 2002.
141. T. I. Bø and T. A. Johansen, "Dynamic safety constraints by scenario based economic model predictive control," *IFAC Proceedings Volumes*, vol. 47, no. 3, pp. 9412–9418, 2014.
142. D. Xue and N. H. El-Farra, "Actuator fault-tolerant control of networked distributed processes with event-triggered sensor-controller communication," in *2016 American Control Conference (ACC)*, pp. 1661–1666, IEEE, 2016.
143. J. T. Allen and N. H. El-Farra, "A model-based framework for fault estimation and accommodation applied to distributed energy resources," *Renewable energy*, vol. 100, pp. 35–43, 2017.
144. Y. Chu and F. You, "Integrated scheduling and dynamic optimization by stackelberg game: bilevel model formulation and efficient solution algorithm," *Industrial & Engineering Chemistry Research*, vol. 53, no. 13, pp. 5564–5581, 2014.
145. M. A. Gutiérrez-Limón, A. Flores-Tlacuahuac, and I. Grossmann, "Scheduling and control using multiobjective optimization approach," *Foundations of Computer-Aided Process Operations Chemical Process Controls*, vol. 11, p. 12, 2012.
146. O. Santander and M. Baldea, "Control-aware batch process scheduling," *Computers & Chemical Engineering*, vol. 152, p. 107360, 2021.
147. E. N. Pistikopoulos and N. A. Diangelakis, "Towards the integration of process design, control and scheduling: Are we getting closer?," *Computers and Chemical Engineering*, vol. 91, pp. 85–92, 2016.
148. S. Misra and R. D. Gudi, "Integration of scheduling & control for sequential batch processes: An iterative approach," *IFAC-PapersOnLine*, vol. 51, no. 1, pp. 84–89, 2018.
149. C. R. Touretzky and M. Baldea, "A hierarchical scheduling and control strategy for thermal energy storage systems," *Energy and Buildings*, vol. 110, pp. 94–107, 2016.
150. M. Baldea, C. R. Touretzky, J. Park, R. C. Pattison, and I. Harjunkoski, "Handling input dynamics in integrated scheduling and control," in *2016 IEEE International Conference on Automation, Quality and Testing, Robotics (AQTR)*, pp. 1–6, IEEE, 2016.
151. M. C. Georgiadis, M. Schenk, E. N. Pistikopoulos, and R. Gani, "The interactions of design control and operability in reactive distillation systems," *Computers & Chemical Engineering*, vol. 26, no. 4-5, pp. 735–746, 2002.
152. G. M. Kopanos and E. N. Pistikopoulos, "Reactive scheduling by a multiparametric programming rolling horizon framework: A case of a network of combined heat and power units," *Industrial and Engineering Chemistry Research*, vol. 53, no. 11, pp. 4366–4386, 2014.
153. J. Zhuge and M. G. Ierapetritou, "Integration of scheduling and control with closed loop implementation," *Industrial and Engineering Chemistry Research*, vol. 51, no. 25, pp. 8550–8565, 2012.

154. K. Frankl, J. Beenken, and W. Marquardt, "Integrated scheduling and control of continuous-time blending processes," in *Computer Aided Chemical Engineering*, vol. 31, pp. 1090–1094, Elsevier, 2012.
155. V. M. Charitopoulos, V. Dua, and L. G. Papageorgiou, "Closed loop integration of planning, scheduling and control via exact multi-parametric nonlinear programming," in *Computer aided chemical engineering*, vol. 40, pp. 1273–1278, Elsevier, 2017.
156. V. M. Charitopoulos, V. Dua, and L. G. Papageorgiou, "Traveling salesman problem-based integration of planning, scheduling, and optimal control for continuous processes," *Industrial & Engineering Chemistry Research*, vol. 56, no. 39, pp. 11186–11205, 2017.
157. O. Andrés-Martínez and L. A. Ricardez-Sandoval, "A nested online scheduling and nonlinear model predictive control framework for multi-product continuous systems," *AIChE Journal*, vol. 68, no. 5, p. e17665, 2022.
158. B. Burnak, N. A. Diangelakis, and E. N. Pistikopoulos, "Process scheduling and control via multiparametric programming," in *Integrated Process Design and Operational Optimization via Multiparametric Programming*, pp. 109–133, Springer, 2020.
159. B. Burnak and E. N. Pistikopoulos, "Integrated process design, scheduling, and model predictive control of batch processes with closed-loop implementation," *AIChE Journal*, vol. 66, no. 10, p. e16981, 2020.
160. B. Burnak, N. A. Diangelakis, and E. N. Pistikopoulos, "Simultaneous process design, scheduling, and advanced model-based control," in *Integrated Process Design and Operational Optimization via Multiparametric Programming*, pp. 135–164, Springer, 2020.
161. L. S. Dias and M. G. Ierapetritou, "Integration of planning, scheduling and control problems using data-driven feasibility analysis and surrogate models," *Computers & Chemical Engineering*, vol. 134, p. 106714, 2020.
162. J. Du, J. Park, I. Harjunkoski, and M. Baldea, "A time scale-bridging approach for integrating production scheduling and process control," *Computers and Chemical Engineering*, vol. 79, pp. 59–69, 2015.
163. M. T. Kelley, R. C. Pattison, R. Baldick, and M. Baldea, "An efficient MILP framework for integrating nonlinear process dynamics and control in optimal production scheduling calculations," *Computers and Chemical Engineering*, vol. 110, pp. 35–52, 2018.
164. J. M. Simkoff and M. Baldea, "Stochastic scheduling and control using data-driven nonlinear dynamic models: application to demand response operation of a chlor-alkali plant," *Industrial & Engineering Chemistry Research*, vol. 59, no. 21, pp. 10031–10042, 2020.
165. A. W. Dowling, T. Zheng, and V. M. Zavala, "A decomposition algorithm for simultaneous scheduling and control of csp systems," *AIChE Journal*, vol. 64, no. 7, pp. 2408–2417, 2018.
166. L. S. Dias and M. G. Ierapetritou, "Integration of scheduling and control under uncertainties: Review and challenges," *Chemical Engineering Research and Design*, vol. 116, pp. 98–113, 2016.
167. H. U. R. Vera and L. A. Ricárdez-Sandoval, "Integration of scheduling and control under stochastic parametric uncertainty with varying unit operation times for chemical batch plants: a back-off approach," *IFAC-PapersOnLine*, vol. 54, no. 3, pp. 578–583, 2021.
168. D. Dering and C. L. Swartz, "A stochastic optimization framework for integrated scheduling and control under demand uncertainty," *Computers & Chemical Engineering*, vol. 165, p. 107931, 2022.
169. Y. I. Valdez-Navarro and L. A. Ricardez-Sandoval, "A novel back-off algorithm for integration of scheduling and control of batch processes under uncertainty," *Industrial & Engineering Chemistry Research*, vol. 58, no. 48, pp. 22064–22083, 2019.
170. R. Al Ismaili, M. W. Lee, D. I. Wilson, and V. S. Vassiliadis, "Heat exchanger network cleaning scheduling: From optimal control to mixed-integer decision making," *Computers & Chemical Engineering*, vol. 111, pp. 1–15, 2018.

171. S. Adloor and V. Vassiliadis, "An optimal control approach to scheduling maintenance and production in parallel lines of reactors using decaying catalysts," *Computers & Chemical Engineering*, vol. 142, p. 107025, 2020.
172. A. Huercio, A. Espuña, and L. Puigjaner, "Incorporating on-line scheduling strategies in integrated batch production control," *Computers and Chemical Engineering*, vol. 19, no. SUPPL. 1, pp. 609–614, 1995.
173. E. Muñoz, E. Capón-García, M. Moreno-Benito, A. Espuña, and L. Puigjaner, "Scheduling and control decision-making under an integrated information environment," *Computers & chemical engineering*, vol. 35, no. 5, pp. 774–786, 2011.
174. L. D. Beal, D. Petersen, G. Pila, B. Davis, S. Warnick, and J. D. Hedengren, "Economic benefit from progressive integration of scheduling and control for continuous chemical processes," *Processes*, vol. 5, no. 4, p. 84, 2017.
175. N. Diangelakis, B. Burnak, and E. Pistikopoulos, "A multi-parametric programming approach for the simultaneous process scheduling and control–application to a domestic cogeneration unit," *Chemical Process Control. Tucson, Arizona, January*, pp. 8–122017, 2017.
176. C. T. Maravelias, *Chemical Production Scheduling: Mixed-Integer Programming Models and Methods*. Cambridge University Press, 2021.
177. D. B. Birewar and I. E. Grossmann, "Incorporating scheduling in the optimal design of multiproduct batch plants," *Computers & Chemical Engineering*, vol. 13, no. 1, pp. 141–161, 1989. Computer Applications to Batch Chemical Processing.
178. V. White, J. Perkins, and D. Espie, "Switchability analysis," *Computers & Chemical Engineering*, vol. 20, no. 4, pp. 469–474, 1996.
179. T. Bhatia and L. Biegler, "Dynamic optimization in the design and scheduling of multiproduct batch plants," *Industrial & engineering chemistry research*, vol. 5885, no. 95, pp. 2234–2246, 1996.
180. T. K. Bhatia and L. T. Biegler, "Dynamic optimization for batch design and scheduling with process model uncertainty," *Industrial and Engineering Chemistry Research*, vol. 36, no. 9, pp. 3708–3717, 1997.
181. M. Baldea and I. Harjunkoski, "Integrated production scheduling and process control: A systematic review," *Computers and Chemical Engineering*, vol. 71, pp. 377–390, 2014.
182. R. Mahadevan, F. J. Doyle III, and A. C. Allcock, "Control-Relevant Scheduling of Polymer Grade Transitions," *AIChE Journal*, vol. 48, no. 8, pp. 1754–1764, 2002.
183. C. Chatzidoukas, J. D. Perkins, E. N. Pistikopoulos, and C. Kiparissides, "Optimal grade transition and selection of closed-loop controllers in a gas-phase olefin polymerization fluidized bed reactor," *Chemical Engineering Science*, vol. 58, no. 0009, pp. 3643-3658, 2003.
184. S. Terrazas-Moreno, A. Flores-Tlacuahuac, and I. E. Grossmann, "Simultaneous cyclic scheduling and optimal control of polymerization reactors," *AIChE Journal*, vol. 53, pp. 2301–2315, 9 2007.
185. A. Flores-Tlacuahuac and I. E. Grossmann, "Simultaneous Cyclic Scheduling and Control of Tubular Reactors: Single Production Lines," *Industrial and Engineering Chemistry Research*, vol. 49, pp. 11453–11463, 2010.
186. A. Flores-Tlacuahuac and I. E. Grossmann, "Simultaneous cyclic scheduling and control of tubular reactors: Parallel production lines," *Industrial and Engineering Chemistry Research*, vol. 50, no. 13, pp. 8086–8096, 2011.
187. M. A. Gutiérrez-Limón, A. Flores-Tlacuahuac, and I. E. Grossmann, "Minlp formulation for simultaneous planning, scheduling, and control of short-period single-unit processing systems," *Industrial & Engineering Chemistry Research*, vol. 53, no. 38, pp. 14679–14694, 2014.
188. C. Loeblein and J. D. Perkins, "Structural design for on-line process optimization: I. dynamic economics of mpc," *AIChE Journal*, vol. 45, no. 5, pp. 1018–1029, 1999.

189. A. C. Zanin, M. T. de Gouvêa, and D. Odloak, "Integrating real-time optimization into the model predictive controller of the fcc system," *Control Engineering Practice*, vol. 10, no. 8, pp. 819–831, 2002.
190. J. B. Rawlings and R. Amrit, *Optimizing Process Economic Performance Using Model Predictive Control*, pp. 119–138. Berlin, Heidelberg: Springer Berlin Heidelberg, 2009.
191. R. Amrit, J. B. Rawlings, and D. Angeli, "Economic optimization using model predictive control with a terminal cost," *Annual Reviews in Control*, vol. 35, no. 2, pp. 178–186, 2011.
192. L. Würth, R. Hannemann, and W. Marquardt, "A two-layer architecture for economically optimal process control and operation," *Journal of Process Control*, vol. 21, no. 3, pp. 311–321, 2011.
193. M. Ellis and P. D. Christofides, "Selection of control configurations for economic model predictive control systems," *AIChE Journal*, vol. 60, no. 9, pp. 3230–3242, 2014.
194. M. Z. Jamaludin and C. L. E. Swartz, "Dynamic real-time optimization with closed-loop prediction," *AIChE Journal*, vol. 63, no. 9, pp. 3896–3911, 2017.
195. H. Li and C. L. E. Swartz, "Dynamic real-time optimization of distributed mpc systems using rigorous closed-loop prediction," *Computers & Chemical Engineering*, vol. 122, pp. 356–371, 2019. 2017 Edition of the European Symposium on Computer Aided Process Engineering (ESCAPE-27).
196. J. Downs and E. Vogel, "A plant-wide industrial process control problem," *Computers & Chemical Engineering*, vol. 17, no. 3, pp. 245–255, 1993. Industrial challenge problems in process control.
197. I. E. Grossmann and R. W. H. Sargent, "Optimum design of heat exchanger networks," *Computers & Chemical Engineering*, vol. 2, no. 1, pp. 1–7, 1978.
198. N. Nishida, G. Stephanopoulos, and A. W. Westerberg, "A review of process synthesis," *AIChE Journal*, vol. 27, pp. 321–351, 5 1981.
199. K. P. Papalexandri and E. N. Pistikopoulos, "Generalized modular representation framework for process synthesis," *AIChE Journal*, vol. 42, pp. 1010–1032, 4 1996.
200. S. E. Demirel, J. Li, and M. M. F. Hasan, "Systematic process intensification using building blocks," *Computers & Chemical Engineering*, vol. 105, pp. 2–38, 2017. Process Intensification.
201. A. K. Tula, D. K. Babi, J. Bottlaender, M. R. Eden, and R. Gani, "A computer-aided software-tool for sustainable process synthesis-intensification," *Computers & Chemical Engineering*, vol. 105, pp. 74–95, 2017. Process Intensification.
202. F. E. da Cruz and V. I. Manousiouthakis, "Process intensification of reactive separator networks through the ideas conceptual framework," *Computers & Chemical Engineering*, vol. 105, pp. 39–55, 2017. Process Intensification.
203. J. Li, S. E. Demirel, and M. M. F. Hasan, "Process synthesis using block superstructure with automated flowsheet generation and optimization," *AIChE Journal*, vol. 64, no. 8, pp. 3082–3100, 2018.
204. J. Li, S. E. Demirel, and M. M. F. Hasan, "Process integration using block superstructure," *Industrial & Engineering Chemistry Research*, vol. 57, no. 12, pp. 4377–4398, 2018.
205. Y. Tian and E. N. Pistikopoulos, "Synthesis of operable process intensification systems—steady-state design with safety and operability considerations," *Industrial & Engineering Chemistry Research*, vol. 58, no. 15, pp. 6049–6068, 2019.
206. S. E. Demirel, J. Li, and M. M. F. Hasan, "Systematic process intensification," *Current Opinion in Chemical Engineering*, 2019.
207. S. E. Demirel, J. Li, and M. M. F. Hasan, "A general framework for process synthesis, integration, and intensification," *Industrial & Engineering Chemistry Research*, vol. 58, no. 15, pp. 5950–5967, 2019.

208. Y. Tian, S. E. Demirel, M. M. F. Hasan, and E. N. Pistikopoulos, "An overview of process systems engineering approaches for process intensification: State of the art," *Chemical Engineering and Processing - Process Intensification*, vol. 133, pp. 160–210, 2018.
209. M. Baldea, "From process integration to process intensification," *Computers & Chemical Engineering*, vol. 81, pp. 104–114, 2015. Special Issue: Selected papers from the 8th International Symposium on the Foundations of Computer-Aided Process Design (FOCAPD 2014), July 13-17, 2014, Cle Elum, Washington, USA.
210. Y. Tian and E. N. Pistikopoulos, "Synthesis of operable process intensification systems: advances and challenges," *Current Opinion in Chemical Engineering*, 2019.
211. L. S. Dias and M. G. Ierapetritou, "Optimal operation and control of intensified processes — challenges and opportunities," *Current Opinion in Chemical Engineering*, 2019.
212. E. N. Pistikopoulos, A. Barbosa-Povoa, J. H. Lee, R. Misener, A. Mitsos, G. V. Reklaitis, V. Venkatasubramanian, F. You, and R. Gani, "Process systems engineering–the generation next?," *Computers & Chemical Engineering*, vol. 147, p. 107252, 2021.
213. I. E. Grossmann, "Advances in mathematical programming models for enterprise-wide optimization," *Computers & Chemical Engineering*, vol. 47, pp. 2–18, 2012.
214. M. Rafiei and L. A. Ricardez-Sandoval, "New frontiers, challenges, and opportunities in integration of design and control for enterprise-wide sustainability," *Computers & Chemical Engineering*, vol. 132, p. 106610, 2020.
215. T. F. Edgar and E. N. Pistikopoulos, "Smart manufacturing and energy systems," *Computers & Chemical Engineering*, vol. 114, pp. 130–144, 2018.
216. T. Coito, B. Firme, M. S. Martins, A. Costigliola, R. Lucas, J. Figueiredo, S. M. Vieira, and J. M. Sousa, "Integration of industrial iot architectures for dynamic scheduling," *Computers & Industrial Engineering*, vol. 171, p. 108387, 2022.
217. J. Zhuge, *Integration of process scheduling and control*. Rutgers The State University of New Jersey, School of Graduate Studies, 2015.
218. J. J. Downs and B. Ogunnaike, "Design for control and operability: An industrial perspective," in *AICHE Symposium Series*, pp. 115–115, AMERICAN INSTITUTE OF CHEMICAL ENGINEERS, 1995.
219. Z. Wu, H. Durand, and P. D. Christofides, "Handling process safety and stochastic uncertainty in economic model predictive control," *IFAC-PapersOnLine*, vol. 51, no. 20, pp. 424–429, 2018.
220. E. N. Pistikopoulos, Y. Tian, and R. Bindlish, "Operability and control in process intensification and modular design: Challenges and opportunities," *AIChE Journal*, vol. 67, no. 5, p. e17204, 2021.
221. Y. Tian and E. N. Pistikopoulos, "Synthesis of operable process intensification systems—steady-state design with safety and operability considerations," *Industrial & Engineering Chemistry Research*, vol. 58, no. 15, pp. 6049–6068, 2018.
222. T. J. Mesquita, G. Campani, R. C. Giordano, M. P. Ribeiro, A. C. Horta, T. C. Zangirolami, and F. V. Lima, "Operability and biomimetic control of a micro-aerated fermentation process," *Computers & Chemical Engineering*, vol. 155, p. 107511, 2021.
223. W. Fisher and J. Douglas, "Analysis of process operability at the preliminary design stage," *Computers & chemical engineering*, vol. 9, no. 5, pp. 499–515, 1985.
224. X. Wang, B. Chen, S. Yang, and C. McGreavy, "Neural nets, fuzzy sets and digraphs in safety and operability studies of refinery reaction processes," *Chemical engineering science*, vol. 51, no. 10, pp. 2169–2178, 1996.
225. E. J. Longwell, "Dynamic modeling for process control and operability," *ISA Transactions*, vol. 33, no. 1, pp. 3–10, 1994.
226. E. M. Al-Mutairi, J. A. Suardin, M. S. Mannan, and M. M. El-Halwagi, "An optimization approach to the integration of inherently safer design and process scheduling," *Journal of Loss Prevention in the Process Industries*, vol. 21, no. 5, pp. 543–549, 2008.

227. E. N. Pistikopoulos, N. A. Diangelakis, R. Oberdieck, M. M. Papathanasiou, I. Nascu, and M. Sun, "PAROC – An integrated framework and software platform for the optimisation and advanced model-based control of process systems," *Chemical Engineering Science*, vol. 136, pp. 115–138, 2015.
228. V. Bansal, V. Sakizlis, R. Ross, J. D. Perkins, and E. N. Pistikopoulos, "New algorithms for mixed-integer dynamic optimization," *Computers and Chemical Engineering*, vol. 27, no. 5, pp. 647–668, 2003.
229. C. Pantelides, "Process modelling in the 1990s," *Fifth World Congress of Chemical Engineering*, 1996. cited By 2.
230. M. Charalambides, N. Shah, and C. Pantelides, "Synthesis of batch reaction/distillation processes using detailed dynamic models," *Computers and Chemical Engineering*, vol. 19, no. SUPPL. 1, pp. 167–174, 1995. cited By 13.
231. H. Furlonge, C. Pantelides, and E. Sørensen, "Optimal operation of multivessel batch distillation columns," *AIChE Journal*, vol. 45, no. 4, pp. 781–801, 1999. cited By 52.
232. T. Ishikawa, Y. Natori, L. Liberis, and C. Pantelides, "Modelling and optimisation of an industrial batch process for the production of dioctyl phthalate," *Computers and Chemical Engineering*, vol. 21, no. SUPPL.1, pp. S1239–S1244, 1997. cited By 11.
233. I. Løvik, M. Hillestad, and T. Hertzberg, "Long term dynamic optimization of a catalytic reactor system," *Computers and Chemical Engineering*, vol. 22, no. SUPPL.1, pp. S707–S710, 1998. cited By 46.
234. M. Georgiadis, G. Rotstein, and S. Macchietto, "Optimal design and operation of heat exchangers under milk fouling," *AIChE Journal*, vol. 44, no. 9, pp. 2099–2111, 1998. cited By 31.
235. S. Viswanathan, N. Shah, and V. Venkatasubramanian, "A hybrid strategy for batch process hazards analysis," *Computers and Chemical Engineering*, vol. 24, no. 2-7, pp. 545–549, 2000. cited By 9.
236. R. Ross, V. Bansal, J. Perkins, E. Pistikopoulos, G. Koot, and J. van Schijndel, "Optimal design and control of an industrial distillation system," *Computers and Chemical Engineering*, vol. 23, no. SUPPL. 1, pp. S875–S878, 1999. cited By 12.
237. L. Biegler and G. Sentoni, "Efficient formulation and solution of nonlinear model predictive control problem," *Latin American Applied Research*, vol. 30, no. 4, pp. 315–324, 2000. cited By 3.
238. R. Nath and Z. Alzein, "On-line dynamic optimization of olefins plants," *Computers and Chemical Engineering*, vol. 24, no. 2-7, pp. 533–538, 2000. cited By 36.
239. Y. Huang, G. Reklaitis, and V. Venkatasubramanian, "Dynamic optimization based fault accommodation," *Computers and Chemical Engineering*, vol. 24, no. 2-7, pp. 439–444, 2000. cited By 20.
240. A. Cervantes, S. Tonelli, A. Brandolin, A. Bandoni, and L. Biegler, "Large-scale dynamic optimization of a low density polyethylene plant," *Computers and Chemical Engineering*, vol. 24, no. 2-7, pp. 983–989, 2000. cited By 22.
241. K. McAuley and J. MacGregor, "Optimal grade transitions in a gas phase polyethylene reactor," *AIChE Journal*, vol. 38, no. 10, pp. 1564–1576, 1992. cited By 133.
242. O. Abel and W. Marquardt, "Scenario-integrated modeling and optimization of dynamic systems," *AIChE Journal*, vol. 46, no. 4, pp. 803–823, 2000. cited By 24.
243. C. Floudas *Nonlinear and Mixed-Integer Optimization: Fundamentals and Applications*, 1995. cited By 1208.
244. R. Sargent and G. Sullivan, "Development of feed changeover policies for refinery distillation units," *Industrial and Engineering Chemistry Process Design and Development*, vol. 18, no. 1, pp. 113–124, 1979. cited By 32.

245. N. Samsatli, L. Papageorgiou, and N. Shah, "Robustness metrics for dynamic optimization models under parameter uncertainty," *AIChE Journal*, vol. 44, no. 9, pp. 1993–2006, 1998. cited By 61.
246. I. Androulakis, "Kinetic mechanism reduction based on an integer programming approach," *AIChE Journal*, vol. 46, no. 2, pp. 361–371, 2000. cited By 80.
247. S. Balakrishna and L. Biegler, "A unified approach for the simultaneous synthesis of reaction, energy, and separation systems," *Industrial and Engineering Chemistry Research*, vol. 32, no. 7, pp. 1372–1382, 1993. cited By 55.
248. M. Avraam, N. Shah, and C. Pantelides, "Modelling and optimisation of general hybrid systems in the continuous time domain," *Computers and Chemical Engineering*, vol. 22, no. SUPPL.1, pp. S221–S228, 1998. cited By 71.
249. M. Avraam, N. Shah, and C. Pantelides, "A decomposition algorithm for the optimisation of hybrid dynamic processes," *Computers and Chemical Engineering*, vol. 23, no. SUPPL. 1, pp. S451–S454, 1999. cited By 17.
250. J. Viswanathan and I. Grossmann, "A combined penalty function and outer approximation method for minlp optimization," *Computers and Chemical Engineering*, vol. 14, no. 7, pp. 769–782, 1990. cited By 556.
251. A. Geoffrion, "Generalized benders decomposition," *Journal of Optimization Theory and Applications*, vol. 10, no. 4, pp. 237–260, 1972. cited By 1349.
252. E. S. Schultz, R. Hannemann-Tamás, and A. Mitsos, "Polynomial approximation of inequality path constraints in dynamic optimization," *Computers & Chemical Engineering*, vol. 135, p. 106732, 2020.
253. V. Vassiliadis, R. Sargent, and C. Pantelides, "Solution of a class of multistage dynamic optimization problems. 2. problems with path constraints," *Industrial and Engineering Chemistry Research*, vol. 33, no. 9, pp. 2123–2133, 1994. cited By 259.
254. V. Vassiliadis, R. Sargent, and C. Pantelides, "Solution of a class of multistage dynamic optimization problems. 1. problems without path constraints," *Industrial and Engineering Chemistry Research*, vol. 33, no. 9, pp. 2111–2122, 1994. cited By 334.
255. R. Pytlak, "Runge-kutta based procedure for the optimal control of differential-algebraic equations," *Journal of Optimization Theory and Applications*, vol. 97, no. 3, pp. 675–705, 1998. cited By 19.
256. K. Bloss, L. Biegler, and W. Schiesser, "Dynamic process optimization through adjoint formulations and constraint aggregation," *Industrial and Engineering Chemistry Research*, vol. 38, no. 2, pp. 421–432, 1999. cited By 37.
257. B. Chachuat, A. B. Singer, and P. I. Barton, "Global methods for dynamic optimization and mixed-integer dynamic optimization," *Industrial & Engineering Chemistry Research*, vol. 45, no. 25, pp. 8373–8392, 2006.
258. A. Cervantes and L. Biegler, "Large-scale dae optimization using a simultaneous nlp formulation," *AIChE Journal*, vol. 44, no. 5, pp. 1038–1050, 1998. cited By 120.
259. R. Allgor and P. Barton, "Mixed-integer dynamic optimization I: problem formulation," *Computers & Chemical Engineering*, vol. 23, no. 4-5, pp. 567–584, 1999.
260. L. Narraway *Selection of Process Control Structure Based on Economics*, 1992. cited By 7.
261. C. Schweiger and C. Floudas, "Interaction of design and control: Optimization with dynamic models," *Optimal Control: Theory, Algorithms and Applications*, pp. 388–435, 1997. cited By 81.
262. M. Sharif, N. Shah, and C. Pantelides, "On the design of multicomponent batch distillation columns," *Computers and Chemical Engineering*, vol. 22, no. SUPPL.1, pp. S69–S76, 1998. cited By 35.

263. M. Duran and I. Grossmann, "An outer-approximation algorithm for a class of mixed-integer nonlinear programs," *Mathematical Programming*, vol. 36, no. 3, pp. 307–339, 1986. cited By 842.
264. M. Mohideen, J. Perkins, and E. Pistikopoulos, "Towards an efficient numerical procedure for mixed integer optimal control," *Computers and Chemical Engineering*, vol. 21, no. SUPPL.1, pp. S457–S462, 1997. cited By 36.
265. A. Bryson and Y. Ho *Applied Optimal Control*, 1975. cited By 7172.
266. C. Floudas and V. Visweswaran, "A global optimization algorithm (gop) for certain classes of nonconvex nlps-i. theory," *Computers and Chemical Engineering*, vol. 14, no. 12, pp. 1397–1417, 1990. cited By 172.
267. N. Sahinidis and I. Grossmann, "Convergence properties of generalized benders decomposition," *Computers and Chemical Engineering*, vol. 15, no. 7, pp. 481–491, 1991. cited By 78.
268. M. Bagajewicz and V. Manousiouthakis, "On the generalized benders decomposition," *Computers and Chemical Engineering*, vol. 15, no. 10, pp. 691–700, 1991. cited By 44.
269. P. Barton, R. Allgor, W. Feehery, and S. Galán, "Dynamic optimization in a discontinuous world," *Industrial and Engineering Chemistry Research*, vol. 37, no. 3, pp. 966–981, 1998. cited By 90.
270. O. Rosen and R. Luus, "Evaluation of gradients for piecewise constant optimal control," *Computers and Chemical Engineering*, vol. 15, no. 4, pp. 273–281, 1991. cited By 48.
271. R. Fletcher and S. Leyffer, "Solving mixed integer nonlinear programs by outer approximation," *Mathematical Programming, Series B*, vol. 66, no. 3, pp. 327–349, 1994. cited By 133.
272. A. Brooke, D. Kendrick, and A. Meeraus *GAMS Release 2.25: A User's Guide*, 1992. cited By 160.
273. M. Duran *A Mixed-Integer Nonlinear Programming Approach for the Systematic Synthesis of Engineering Systems*, 1984. cited By 8.
274. P. J. Campo and M. Morari, "Robust model predictive control.," pp. 1021–1026, 1987. cited By 295.
275. J. Lee and Z. Yu, "Worst-case formulations of model predictive control for systems with bounded parameters," *Automatica*, vol. 33, no. 5, pp. 763–781, 1997. cited By 314.
276. A. Schwarm and M. Nikolaou, "Chance-constrained model predictive control," *AIChE Journal*, vol. 45, no. 8, pp. 1743–1752, 1999. cited By 175.
277. T. Badgwell, "Robust model predictive control of stable linear systems," *International Journal of Control*, vol. 68, no. 4, pp. 797–818, 1997. cited By 100.
278. D. Kassmann, T. Badgwell, and R. Hawkins, "Robust steady-state target calculation for model predictive control," *AIChE Journal*, vol. 46, no. 5, pp. 1007–1024, 2000. cited By 129.
279. P. Scokaert and D. Mayne, "Min-max feedback model predictive control for constrained linear systems," *IEEE Transactions on Automatic Control*, vol. 43, no. 8, pp. 1136–1142, 1998. cited By 580.
280. J. Lee and B. Cooley, "Min-max predictive control techniques for a linear state-space system with a bounded set of input matrices," *Automatica*, vol. 36, no. 3, pp. 463–473, 2000. cited By 47.
281. D. Mayne and W. Schroeder, "Robust time-optimal control of constrained linear systems," *Automatica*, vol. 33, no. 12, pp. 2103–2118, 1997. cited By 131.
282. A. Bemporad, F. Borrelli, and M. Morari, "Min-max control of constrained uncertain discrete-time linear systems," *IEEE Transactions on Automatic Control*, vol. 48, no. 9, pp. 1600–1606, 2003. cited By 352.
283. H. Khajuria and E. Pistikopoulos, "Dynamic modeling and explicit/multi-parametric mpc control of pressure swing adsorption systems," *Journal of Process Control*, vol. 21, no. 1, pp. 151–163, 2011. cited By 35.

284. A. Krieger, N. Panoskaltsis, A. Mantalaris, M. Georgiadis, and E. Pistikopoulos, "Modeling and analysis of individualized pharmacokinetics and pharmacodynamics for volatile anesthesia," *IEEE Transactions on Biomedical Engineering*, vol. 61, no. 1, pp. 25–34, 2014. cited By 17.
285. A. Saltelli, K. Chan, and M. Scott *Sensitivity Analysis, Probability and Statistics Series*, 2000. cited By 202.
286. C. Kontoravdi, S. Asprey, E. Pistikopoulos, and A. Mantalaris, "Application of global sensitivity analysis to determine goals for design of experiments: An example study on antibody-producing cell cultures," *Biotechnology Progress*, vol. 21, no. 4, pp. 1128–1135, 2005. cited By 61.
287. A. Kiparissides, S. Kucherenko, A. Mantalaris, and E. Pistikopoulos, "Global sensitivity analysis challenges in biological systems modeling," *Industrial and Engineering Chemistry Research*, vol. 48, no. 15, pp. 7168–7180, 2009. cited By 90.
288. I. Nascu, R. Oberdieck, and E. N. Pistikopoulos, "Explicit hybrid model predictive control strategies for intravenous anaesthesia," *Computers & Chemical Engineering*, 2016.
289. R. S. Lambert, P. Rivotti, and E. N. Pistikopoulos, "A monte-carlo based model approximation technique for linear model predictive control of nonlinear systems," *Computers & Chemical Engineering*, vol. 54, no. 0, pp. 60–67, 2013.
290. P. Van Overschee and B. De Moor, "N4SID: Subspace algorithms for the identification of combined deterministic-stochastic systems," *Automatica*, vol. 30, no. 1, pp. 75–93, 1994.
291. L. Ljung, "System identification - theory for the user," in *Practice Hall Information and System Sciences* (T. Kailath, ed.), New Jersey: Prentice-Hall PTR, 1987.
292. F. Boukouvala and C. A. Floudas, "ARGONAUT: AlgoRithms for Global Optimization of coNstrAined grey-box compUTational problems," *Optimization Letters*, pp. 1–19, 2016.
293. F. Boukouvala, R. Misener, and C. A. Floudas, "Global optimization advances in mixed-integer nonlinear programming, minlp, and constrained derivative-free optimization, cdfo," *European Journal of Operational Research*, vol. 252, no. 3, pp. 701–727, 2016.
294. J.-A. Müller and F. Lemke, "Self-organizing modelling and decision support in economics," *System Analysis Modeling Simulation*, vol. 18-19, pp. 135–138, 1995. cited By 3.
295. I. Sobol, "Global sensitivity indices for nonlinear mathematical models and their monte carlo estimates," *Mathematics and Computers in Simulation*, vol. 55, no. 1-3, pp. 271–280, 2001. cited By 1940.
296. C. Kontoravdi, E. Pistikopoulos, and A. Mantalaris, "Systematic development of predictive mathematical models for animal cell cultures," *Computers and Chemical Engineering*, vol. 34, no. 8, pp. 1192–1198, 2010. cited By 36.
297. T. Homma and A. Saltelli, "Importance measures in global sensitivity analysis of nonlinear models," *Reliability Engineering and System Safety*, vol. 52, no. 1, pp. 1–17, 1996. cited By 942.
298. A. Saltelli, P. Annoni, I. Azzini, F. Campolongo, M. Ratto, and S. Tarantola, "Variance based sensitivity analysis of model output. design and estimator for the total sensitivity index," *Computer Physics Communications*, vol. 181, no. 2, pp. 259–270, 2010. cited By 1006.
299. P. Rivotti, R. Lambert, and E. Pistikopoulos, "Combined model approximation techniques and multiparametric programming for explicit nonlinear model predictive control," *Computers and Chemical Engineering*, vol. 42, pp. 277–287, 2012. cited By 26.
300. J. Hahn and T. Edgar, "An improved method for nonlinear model reduction using balancing of empirical gramians," *Computers and Chemical Engineering*, vol. 26, no. 10, pp. 1379–1397, 2002. cited By 156.
301. E. N. Pistikopoulos, "Perspectives in multiparametric programming and explicit model predictive control," *AIChE Journal*, vol. 55, no. 8, pp. 1918–1925, 2009.
302. T. Gal and J. Nedoma, "Multiparametric linear programming," *Management Science*, vol. 18, no. 7, pp. 406–422, 1972. cited By 154.

303. T. Gal and G. Davis *Postoptimal Analyses, Parametric Programming, and Related Topics*, 1978. cited By 382.
304. A. Bemporad, F. Borrelli, and M. Morari, "Piecewise linear optimal controllers for hybrid systems," vol. 2, pp. 1190–1194, 2000. cited By 141.
305. V. Dua, N. Bozinis, and E. Pistikopoulos, "A multiparametric programming approach for mixed-integer quadratic engineering problems," *Computers and Chemical Engineering*, vol. 26, no. 4-5, pp. 715–733, 2002. cited By 170.
306. V. Dua, K. Papalexandri, and E. Pistikopoulos, "A parametric mixed-integer global optimization framework for the solution of process engineering problems under uncertainty," *Computers and Chemical Engineering*, vol. 23, no. SUPPL. 1, pp. S19–S22, 1999. cited By 12.
307. V. Dua, K. Papalexandri, and E. Pistikopoulos, "Global optimization issues in multiparametric continuous and mixed-integer optimization problems," *Journal of Global Optimization*, vol. 30, no. 1, pp. 59–89, 2004. cited By 61.
308. A. Fiacco and J. Kyparisis, "Convexity and concavity properties of the optimal value function in parametric nonlinear programming," *Journal of Optimization Theory and Applications*, vol. 48, no. 1, pp. 95–126, 1986. cited By 54.
309. E. Zafiriou, "Robust model predictive control of processes with hard constraints," *Computers and Chemical Engineering*, vol. 14, no. 4-5, pp. 359–371, 1990. cited By 127.
310. A. Fiacco and Y. Ishizuka, "Sensitivity and stability analysis for nonlinear programming," *Annals of Operations Research*, vol. 27, no. 1, pp. 215–235, 1990. cited By 57.
311. H. Benson, "Algorithms for parametric nonconvex programming," *Journal of Optimization Theory and Applications*, vol. 38, no. 3, pp. 319–340, 1982. cited By 21.
312. A. Bemporad, F. Borrelli, and M. Morari, "Model predictive control based on linear programming - the explicit solution," *IEEE Transactions on Automatic Control*, vol. 47, no. 12, pp. 1974–1985, 2002. cited By 478.
313. F. Borrelli, M. Baotić, A. Bemporad, and M. Morari, "Dynamic programming for constrained optimal control of discrete-time linear hybrid systems," *Automatica*, vol. 41, no. 10, pp. 1709–1721, 2005. cited By 233.
314. R. Khalilpour and I. Karimi, "Parametric optimization with uncertainty on the left hand side of linear programs," *Computers and Chemical Engineering*, vol. 60, pp. 31–40, 2014. cited By 9.
315. P. Tøndel, T. Johansen, and A. Bemporad, "An algorithm for multi-parametric quadratic programming and explicit mpc solutions," *Automatica*, vol. 39, no. 3, pp. 489–497, 2003. cited By 357.
316. D. Mayne and S. Raković, "Optimal control of constrained piecewise affine discrete-time systems," *Computational Optimization and Applications*, vol. 25, no. 1-3, pp. 167–191, 2003. cited By 16.
317. J. Spjøtvold, E. Kerrigan, C. Jones, P. Tøndel, and T. Johansen, "On the facet-to-facet property of solutions to convex parametric quadratic programs," *Automatica*, vol. 42, no. 12, pp. 2209–2214, 2006. cited By 51.
318. C. Feller and T. Johansen, "Explicit mpc of higher-order linear processes via combinatorial multi-parametric quadratic programming," pp. 536–541, 2013. cited By 9.
319. J. Acevedo and E. Pistikopoulos, "A multiparametric programming approach for linear process engineering problems under uncertainty," *Industrial and Engineering Chemistry Research*, vol. 36, no. 3, pp. 717–728, 1997. cited By 122.
320. Z. Li and M. Ierapetritou, "Process scheduling under uncertainty using multiparametric programming," *AIChE Journal*, vol. 53, no. 12, pp. 3183–3203, 2007. cited By 39.
321. V. Dua and E. Pistikopoulos, "An algorithm for the solution of multiparametric mixed integer linear programming problems," *Annals of Operations Research*, vol. 99, no. 1-4, pp. 123–139, 2000. cited By 171.

322. M. Wittmann-Hohlbein and E. Pistikopoulos, "A two-stage method for the approximate solution of general multiparametric mixed-integer linear programming problems," *Industrial and Engineering Chemistry Research*, vol. 51, no. 23, pp. 8095–8107, 2012. cited By 14.
323. P. Rivotti and E. Pistikopoulos, "Constrained dynamic programming of mixed-integer linear problems by multi-parametric programming," *Computers and Chemical Engineering*, vol. 70, pp. 172–179, 2014. cited By 10.
324. R. Oberdieck, M. Wittmann-Hohlbein, and E. Pistikopoulos, "A branch and bound method for the solution of multiparametric mixed integer linear programming problems," *Journal of Global Optimization*, vol. 59, no. 2-3, pp. 527–543, 2014. cited By 20.
325. V. M. Charitopoulos and V. Dua, "Explicit Model Predictive Control of Hybrid Systems and Multiparametric Mixed Integer Polynomial Programming," *AIChE Journal*, vol. 62, no. 9, pp. 3441–3460, 2016.
326. S. Almér and M. Morari, "Efficient online solution of multi-parametric mixed-integer quadratic problems," *International Journal of Control*, vol. 86, no. 8, pp. 1386–1396, 2013. cited By 4.
327. D. Axehill, T. Besselmann, D. Raimondo, and M. Morari, "A parametric branch and bound approach to suboptimal explicit hybrid mpc," *Automatica*, vol. 50, no. 1, pp. 240–246, 2014. cited By 24.
328. Y. Han and Z. Chen, "Continuity of parametric mixed-integer quadratic programs and its application to stability analysis of two-stage quadratic stochastic programs with mixed-integer recourse," *Optimization*, vol. 64, no. 9, pp. 1983–1997, 2015.
329. M. Seron, G. Goodwin, and J. De Doná, "Finitely parameterised implementation of receding horizon control for constrained linear systems," *Proceedings of the American Control Conference*, vol. 6, pp. 4481–4486, 2002. cited By 11.
330. J. Acevedo and E. Pistikopoulos, "Algorithm for multiparametric mixed-integer linear programming problems," *Operations Research Letters*, vol. 24, no. 3, pp. 139–148, 1999. cited By 40.
331. A. Crema, "An algorithm to perform a complete parametric analysis relative to the constraint matrix for a 0 1-integer linear program," *European Journal of Operational Research*, vol. 138, no. 3, pp. 484–494, 2002. cited By 2.
332. A. Pertsinidis, I. Grossmann, and G. McRae, "Parametric optimization of milp programs and a framework for the parametric optimization of minlps," *Computers and Chemical Engineering*, vol. 22, no. SUPPL.1, pp. S205–S212, 1998. cited By 75.
333. K. Papalexandri and T. Dimkou, "A parametric mixed-integer optimization algorithm for multiobjective engineering problems involving discrete decisions," *Industrial and Engineering Chemistry Research*, vol. 37, no. 5, pp. 1866–1882, 1998. cited By 53.
334. R. Oberdieck, N. Diangelakis, I. Nascu, M. Papathanasiou, M. Sun, S. Avraamidou, and E. Pistikopoulos, "On multi-parametric programming and its applications in process systems engineering," *Chemical Engineering Research and Design*, vol. 116, pp. 61–82, 2016.
335. M. Wittmann-Hohlbein and E. Pistikopoulos, "On the global solution of multi-parametric mixed integer linear programming problems," *Journal of Global Optimization*, vol. 57, no. 1, pp. 51–73, 2013. cited By 17.
336. M. Wittmann-Hohlbein and E. Pistikopoulos, "Approximate solution of mp-milp problems using piecewise affine relaxation of bilinear terms," *Computers and Chemical Engineering*, vol. 61, pp. 136–155, 2014. cited By 6.
337. P. Rivotti and E. Pistikopoulos, "A dynamic programming based approach for explicit model predictive control of hybrid systems," *Computers and Chemical Engineering*, vol. 72, pp. 126–144, 2015. cited By 20.
338. Z. Gajic and M. Qureshi *Lyapunov Matrix Equation in System Stability and Control*, 1995. cited By 289.

339. L. Biegler, A. Cervantes, and A. Wächter, "Advances in simultaneous strategies for dynamic process optimization," *Chemical Engineering Science*, vol. 57, no. 4, pp. 575–593, 2002. cited By 313.
340. S. Kameswaran and L. Biegler, "Convergence rates for direct transcription of optimal control problems using collocation at radau points," *Computational Optimization and Applications*, vol. 41, no. 1, pp. 81–126, 2008. cited By 143.
341. E. N. Pistikopoulos, L. Dominguez, C. Panos, K. Kouramas, and A. Chinchuluun, "Theoretical and algorithmic advances in multi-parametric programming and control," *Computational Management Science*, vol. 9, no. 2, pp. 183–203, 2012.
342. E. Kreindler, "Additional necessary conditions for optimal control with state-variable inequality constraints," *Journal of Optimization Theory and Applications*, vol. 38, no. 2, pp. 241–250, 1982. cited By 35.
343. R. Vinter and H. Zheng, "Necessary conditions for optimal control problems with state constraints," *Transactions of the American Mathematical Society*, vol. 350, no. 3, pp. 1181–1204, 1998. cited By 31.
344. K. Malanowski and H. Maurer, "Sensitivity analysis for optimal control problems subject to higher order state constraints," *Annals of Operations Research*, vol. 101, no. 1-4, pp. 43–73, 2001. cited By 47.
345. D. Augustin and H. Maurer, "Computational sensitivity analysis for state constrained optimal control problems," *Annals of Operations Research*, vol. 101, no. 1-4, pp. 75–99, 2001. cited By 21.
346. V. Sakizlis, J. Perkins, and E. Pistikopoulos, "Explicit solutions to optimal control problems for constrained continuous-time linear systems," vol. 152, pp. 443–452, 2005. cited By 25.
347. A. Polyanin and V. Zaitsev *Handbook of Exact Solutions for Ordinary Differential Equations*, 1995. cited By 1056.
348. J. Solís-Daun, J. Álvarez Ramírez, and R. Suárez, "Semiglobal stabilization of linear systems using constrained controls: A parametric optimization approach," *International Journal of Robust and Nonlinear Control*, vol. 9, no. 8, pp. 461–484, 1999. cited By 7.
349. A. Bemporad and M. Morari, "Control of systems integrating logic, dynamics, and constraints," *Automatica*, vol. 35, no. 3, pp. 407–427, 1999. cited By 2085.
350. A. Dontchev, W. Hager, and K. Malanowski, "Error bounds for euler approximation of a state and control constrained optimal control problem," *Numerical Functional Analysis and Optimization*, vol. 21, no. 5, pp. 653–682, 2000. cited By 39.
351. M. Diehl, H. Bock, J. Schlöder, R. Findeisen, Z. Nagy, and F. Allgöwer, "Real-time optimization and nonlinear model predictive control of processes governed by differential-algebraic equations," *Journal of Process Control*, vol. 12, no. 4, pp. 577–585, 2002. cited By 435.
352. N. Faísca, V. Dua, B. Rustem, P. Saraiva, and E. Pistikopoulos, "Parametric global optimisation for bilevel programming," *Journal of Global Optimization*, vol. 38, no. 4, pp. 609–623, 2007. cited By 72.
353. S. Mariethoz, A. Domahidi, and M. Morari, "High-bandwidth explicit model predictive control of electrical drives," *IEEE Transactions on Industry Applications*, vol. 48, no. 6, pp. 1980–1992, 2012. cited By 74.
354. P. Dua, F. Doyle III, and E. Pistikopoulos, "Model-based blood glucose control for type 1 diabetes via parametric programming," *IEEE Transactions on Biomedical Engineering*, vol. 53, no. 8, pp. 1478–1491, 2006. cited By 139.
355. W. Heemels, B. De Schutter, and A. Bemporad, "Equivalence of hybrid dynamical models," *Automatica*, vol. 37, no. 7, pp. 1085–1091, 2001. cited By 541.
356. A. Ben-Tal and A. Nemirovski, "Robust solutions of linear programming problems contaminated with uncertain data," *Mathematical Programming, Series B*, vol. 88, no. 3, pp. 411–424, 2000. cited By 923.

357. V. Sakizlis, N. Kakalis, V. Dua, J. Perkins, and E. Pistikopoulos, "Design of robust model-based controllers via parametric programming," *Automatica*, vol. 40, no. 2, pp. 189–201, 2004. cited By 84.
358. T. Alamo, D. Ramírez, and E. Camacho, "Efficient implementation of constrained min-max model predictive control with bounded uncertainties: A vertex rejection approach," *Journal of Process Control*, vol. 15, no. 2, pp. 149–158, 2005. cited By 32.
359. S. Olaru and P. Ayerbe, "Robustification of explicit predictive control laws," pp. 4556–4561, 2006. cited By 14.
360. E. N. Pistikopoulos, N. P. Faísca, K. I. Kouramas, and C. Panos, "Explicit robust model predictive control," *Proceedings of the international symposium on advanced control of chemical processes*, vol. ADCHEM'09, pp. 249–254, 2009.
361. E. Kerrigan and J. Maciejowski, "Feedback min-max model predictive control using a single linear program: Robust stability and the explicit solution," *International Journal of Robust and Nonlinear Control*, vol. 14, no. 4, pp. 395–413, 2004. cited By 150.
362. K. I. Kouramas, C. Panos, N. P. Faísca, and E. N. Pistikopoulos, "An algorithm for robust explicit/multi-parametric model predictive control," *Automatica*, vol. 49, no. 2, pp. 381–389, 2013.
363. J.-H. Ryu, V. Dua, and E. Pistikopoulos, "Proactive scheduling under uncertainty: A parametric optimization approach," *Industrial and Engineering Chemistry Research*, vol. 46, no. 24, pp. 8044–8049, 2007. cited By 31.
364. J.-H. Ryu and E. Pistikopoulos, "A novel approach to scheduling of zero-wait batch processes under processing time variations," *Computers and Chemical Engineering*, vol. 31, no. 3, pp. 101–106, 2007. cited By 28.
365. Z. Li and M. Ierapetritou, "Reactive scheduling using parametric programming," *AIChE Journal*, vol. 54, no. 10, pp. 2610–2623, 2008. cited By 42.
366. H. Khajuria and E. Pistikopoulos, "Optimization and control of pressure swing adsorption processes under uncertainty," *AIChE Journal*, vol. 59, no. 1, pp. 120–131, 2013. cited By 49.
367. E. Velliou, M. Fuentes-Garí, R. Misener, E. Pefani, M. Rende, N. Panoskaltsis, A. Mantalaris, and E. Pistikopoulos, "A framework for the design, modeling and optimization of biomedical systems," *Computer Aided Chemical Engineering*, vol. 34, pp. 225–236, 2014. cited By 14.
368. H. Chang, A. Krieger, A. Astolfi, and E. Pistikopoulos, "Robust multi-parametric model predictive control for lpv systems with application to anaesthesia," *Journal of Process Control*, vol. 24, no. 10, pp. 1538–1547, 2014. cited By 11.
369. D. Mayne, J. Rawlings, C. Rao, and P. Scokaert, "Constrained model predictive control: Stability and optimality," *Automatica*, vol. 36, no. 6, pp. 789–814, 2000. cited By 5301.
370. M. Tenny, "Computational strategies for nonlinear model predictive control," *Computational Strategies for Nonlinear Model Predictive Control*, 2002. cited By 19.
371. M. Darby and M. Nikolaou, "A parametric programming approach to moving-horizon state estimation," *Automatica*, vol. 43, no. 5, pp. 885–891, 2007. cited By 40.
372. A. Voelker, K. Kouramas, and E. N. Pistikopoulos, "Simultaneous constrained moving horizon state estimation and model predictive control by multi-parametric programming," in *49th IEEE Conference on Decision and Control*, pp. 5019–5024, 2010.
373. A. Voelker, K. Kouramas, and E. N. Pistikopoulos, "Simultaneous constrained moving horizon state estimation and model predictive control by multi-parametric programming," in *49th IEEE Conference on Decision and Control (CDC)*, pp. 5019–5024, IEEE, 2010.
374. R. Lambert, I. Nascu, and E. N. Pistikopoulos, "Simultaneous reduced order multi-parametric moving horizon estimation and model predictive control," in *Dynamics and Control of Process Systems*, IFAC proceedings volumes, pp. 45–50, Elsevier, 2013.

375. S. Zavitsanou, A. Mantalaris, M. C. Georgiadis, and E. N. Pistikopoulos, “Optimization of insulin dosing in patients with type 1 diabetes mellitus,” in *Computer Aided Chemical Engineering*, vol. 33, pp. 1459–1464, Elsevier, 2014.
376. R. Oberdieck, N. A. Diangelakis, M. M. Papathanasiou, I. Nascu, and E. N. Pistikopoulos, “POP - Parametric Optimization Toolbox,” *Industrial and Engineering Chemistry Research*, vol. 55, no. 33, pp. 8979–8991, 2016.
377. N. Diangelakis, C. Panos, and E. Pistikopoulos, “Design optimization of an internal combustion engine powered chp system for residential scale application,” *Computational Management Science*, vol. 11, no. 3, pp. 237–266, 2014.
378. H. Li, F. Marechal, and D. Favrat, “Power and cogeneration technology environomic performance typification in the context of co<inf>2</inf> abatement part ii: Combined heat and power cogeneration,” *Energy*, vol. 35, no. 9, pp. 3517–3523, 2010. cited By 32.
379. D. Wu and R. Wang, “Combined cooling, heating and power: A review,” *Progress in Energy and Combustion Science*, vol. 32, no. 5-6, pp. 459–495, 2006. cited By 583.
380. D. Konstantinidis, P. Varbanov, and J. Klemeš, “Multi-parametric control and optimisation of a small scale chp,” *Chemical Engineering Transactions*, vol. 21, pp. 151–156, 2010. cited By 9.
381. L. Guzzella and C. Onder, *Introduction to modeling and control of internal combustion engine systems*. 2010. cited By 798.
382. J. Heywood *Internal Combustion Engine Fundamentals*, 1988. cited By 14723.
383. C. Ziogou, E. Pistikopoulos, M. Georgiadis, S. Voutetakis, and S. Papadopoulou, “Empowering the performance of advanced nmpc by multiparametric programming - an application to a pem fuel cell system,” *Industrial and Engineering Chemistry Research*, vol. 52, no. 13, pp. 4863–4873, 2013. cited By 10.
384. C. Ziogou, M. Georgiadis, E. Pistikopoulos, S. Papadopoulou, and S. Voutetakis, “Combining multi-parametric programming and nmpc for the efficient operation of a pem fuel cell,” *Chemical Engineering Transactions*, vol. 35, pp. 913–918, 2013. cited By 1.
385. Y. Tian, I. Pappas, B. Burnak, J. Katz, and E. N. Pistikopoulos, “A systematic framework for the synthesis of operable process intensification systems – reactive separation systems,” *Computers & Chemical Engineering*, vol. 134, p. 106675, 2020.
386. L. A. Ricardez-Sandoval, “Optimal design and control of dynamic systems under uncertainty: A probabilistic approach,” *Computers & Chemical Engineering*, vol. 43, pp. 91–107, 2012.
387. A. Flores-Tlacuahuac and L. Biegler, “Simultaneous mixed-integer dynamic optimization for integrated design and control,” *Computers and Chemical Engineering*, vol. 31, no. 5-6, pp. 588–600, 2007.
388. M. Alvarado-Morales, M. Hamid, G. Sin, K. Gernaey, J. Woodley, and R. Gani, “A model-based methodology for simultaneous design and control of a bioethanol production process,” *Computers and Chemical Engineering*, vol. 34, no. 12, pp. 2043–2061, 2010.
389. S. Skogestad and M. Morari, “Control configuration selection for distillation columns,” *AIChE Journal*, vol. 33, no. 10, pp. 1620–1635, 1987.
390. M. Soliman, C. Swartz, and R. Baker, “A mixed-integer formulation for back-off under constrained predictive control,” *Computers and Chemical Engineering*, vol. 32, no. 10, pp. 2409–2419, 2008.
391. M. Francisco, P. Vega, and H. Álvarez, “Robust integrated design of processes with terminal penalty model predictive controllers,” *Chemical Engineering Research and Design*, vol. 89, no. 7, pp. 1011–1024, 2011.
392. K. I. Kouramas, N. P. Faísca, C. Panos, and E. N. Pistikopoulos, “Explicit/multi-parametric model predictive control (mpc) of linear discrete-time systems by dynamic and multi-parametric programming,” *Automatica*, vol. 47, no. 8, pp. 1638–1645, 2011.
393. I. Bogle and M. Rashid, “An assessment of dynamic operability measures,” *Computers and Chemical Engineering*, vol. 13, no. 11-12, pp. 1277–1282, 1989.

394. E. Fraga, J. Hagemann, A. Estrada-Villagrana, and I. Bogle, "Incorporation of dynamic behaviour in an automated process synthesis system," *Computers and Chemical Engineering*, vol. 24, no. 2-7, pp. 189–194, 2000.
395. K. Papalexandri and E. Pistikopoulos, "Synthesis and retrofit design of operable heat exchanger networks. 1.flexibility and structural controllability aspects," *Industrial and Engineering Chemistry Research*, vol. 33, no. 7, pp. 1718–1737, 1994.
396. M. Georgiadis, M. Schenk, R. Gani, and E. Pistikopoulos, "The interactions of design, control and operability in reactive distillation systems," *Computer Aided Chemical Engineering*, vol. 9, no. C, pp. 997–1002, 2001.
397. M. Luyben and C. Floudas, "Analyzing the interaction of design and control-1. a multiobjective framework and application to binary distillation synthesis," *Computers and Chemical Engineering*, vol. 18, no. 10, pp. 933–969, 1994.
398. C. Floudas, "Global optimization in design and control of chemical process systems," *Journal of Process Control*, vol. 10, no. 2, pp. 125–134, 2000.
399. C. Floudas, Z. Gümüş, and M. Ierapetritou, "Global optimization in design under uncertainty: Feasibility test and flexibility index problems," *Industrial and Engineering Chemistry Research*, vol. 40, no. 20, pp. 4267–4282, 2001.
400. P. Bahri, J. Bandoni, and J. Romagnoli, "Integrated flexibility and controllability analysis in design of chemical processes," *AIChE Journal*, vol. 43, no. 4, pp. 997–1015, 1997.
401. N. Chawankul, L. Ricardez Sandoval, H. Budman, and P. Douglas, "Integration of design and control: A robust control approach using mpc," *Canadian Journal of Chemical Engineering*, vol. 85, no. 4, pp. 433–446, 2007.
402. K. Sánchez-Sánchez and L. Ricardez-Sandoval, "Simultaneous process synthesis and control design under uncertainty: A worst-case performance approach," *AIChE Journal*, vol. 59, no. 7, pp. 2497–2514, 2013.
403. W. R. Fisher, M. F. Doherty, and J. M. Douglas, "Interface between design and control. 1. process controllability.," *Industrial and Engineering Chemistry Research*, vol. 27, no. 4, pp. 597–605, 1988.
404. W. R. Fisher, M. F. Doherty, and J. M. Douglas, "Interface between design and control. 2. process operability.," *Industrial and Engineering Chemistry Research*, vol. 27, no. 4, pp. 606–611, 1988.
405. W. R. Fisher, M. F. Doherty, and J. M. Douglas, "Interface between design and control. 3. selecting a set of controlled variables.," *Industrial and Engineering Chemistry Research*, vol. 27, no. 4, pp. 611–615, 1988.
406. S. Skogestad and M. Morari, "Design of resilient processing plants-ix. effect of model uncertainty on dynamic resilience," *Chemical Engineering Science*, vol. 42, no. 7, pp. 1765–1780, 1987.
407. N. Chatrattanawet, S. Skogestad, and A. Arpornwichanop, "Control structure design and controllability analysis for solid oxide fuel cell," *Chemical Engineering Transactions*, vol. 39, no. Special Issue, pp. 1291–1296, 2014.
408. E. Close, J. Salm, D. Bracewell, and E. Sorensen, "A model based approach for identifying robust operating conditions for industrial chromatography with process variability," *Chemical Engineering Science*, vol. 116, pp. 284–295, 2014.
409. M. Gevelber and G. Stephanopoulos, "Control and system design for the czochralski crystal growth process," in *American Society of Mechanical Engineers, Dynamic Systems and Control Division (Publication) DSC*, vol. 9, pp. 35–40, 1988.
410. I. Banerjee and M. Ierapetritou, "Design optimization under parameter uncertainty for general black-box models," *Industrial and Engineering Chemistry Research*, vol. 41, no. 26, pp. 6687–6697, 2002.

411. J. Gong, G. Hytoft, and R. Gani, "An integrated computer aided system for integrated design and control of chemical processes," *Computers and Chemical Engineering*, vol. 19, no. SUPPL. 1, pp. 489–494, 1995.
412. J. Figueroa, P. Bahri, J. Bandoni, and J. Romagnoli, "Economic impact of disturbances and uncertain parameters in chemical processes - a dynamic back-off analysis," *Computers and Chemical Engineering*, vol. 20, no. 4, pp. 453–461, 1996.
413. P. Vega, R. Lamanna, S. Revollar, and M. Francisco, "Integrated design and control of chemical processes - part ii: An illustrative example," *Computers and Chemical Engineering*, vol. 71, pp. 618–635, 2014.
414. M. Soroush and C. Kravaris, "Optimal design and operation of batch reactors. 2. a case study," *Industrial and Engineering Chemistry Research*, vol. 32, no. 5, pp. 882–893, 1993.
415. M. Soroush and C. Kravaris, "Optimal design and operation of batch reactors. 1. theoretical framework," *Industrial and Engineering Chemistry Research*, vol. 32, no. 5, pp. 866–881, 1993.
416. V. Bansal, R. Ross, J. Perkins, and E. Pistikopoulos, "Interactions of design and control: Double-effect distillation," *Journal of Process Control*, vol. 10, no. 2, pp. 219–227, 2000.
417. V. Bansal, J. Perkins, and E. Pistikopoulos, "A case study in simultaneous design and control using rigorous, mixed-integer dynamic optimization models," *Industrial and Engineering Chemistry Research*, vol. 41, no. 4, pp. 760–778, 2002.
418. M. Georgiadis, M. Schenk, E. Pistikopoulos, and R. Gani, "The interactions of design, control and operability in reactive distillation systems," *Computers and Chemical Engineering*, vol. 26, no. 4-5, pp. 735–746, 2002.
419. I. Washington and C. Swartz, "Design under uncertainty using parallel multiperiod dynamic optimization," *AIChE Journal*, vol. 60, no. 9, pp. 3151–3168, 2014.
420. V. Sakizlis, J. Perkins, and E. Pistikopoulos, "Parametric controllers in simultaneous process and control design," *Computer Aided Chemical Engineering*, vol. 15, no. C, pp. 1020–1025, 2003.
421. V. Sakizlis, J. Perkins, and E. Pistikopoulos, "Recent advances in optimization-based simultaneous process and control design," *Computers and Chemical Engineering*, vol. 28, no. 10, pp. 2069–2086, 2004.
422. V. Sakizlis, J. Perkins, and E. Pistikopoulos, "Chapter b1 simultaneous process and control design using mixed integer dynamic optimization and parametric programming," *Computer Aided Chemical Engineering*, vol. 17, no. C, pp. 187–215, 2004.
423. C. Noeres, K. Dadhe, R. Gesthuisen, S. Engell, and A. Górak, "Model-based design, control and optimisation of catalytic distillation processes," *Chemical Engineering and Processing: Process Intensification*, vol. 43, no. 3, pp. 421–434, 2004.
424. A. Flores-Tlacuahuac and L. Biegler, "Integrated control and process design during optimal polymer grade transition operations," *Computers and Chemical Engineering*, vol. 32, no. 11, pp. 2823–2837, 2008.
425. S. Mehta and L. Ricardez-Sandoval, "Integration of design and control of dynamic systems under uncertainty: A new back-off approach," *Industrial and Engineering Chemistry Research*, vol. 55, no. 2, pp. 485–498, 2016.
426. I. Kookos and J. Perkins, "An algorithmic method for the selection of multivariable process control structures," *Journal of Process Control*, vol. 12, no. 1, pp. 85–99, 2002.
427. I. Kookos and J. Perkins, "Chapter b2 the back-off approach to simultaneous design and control," *Computer Aided Chemical Engineering*, vol. 17, no. C, pp. 216–238, 2004.
428. R.-N. De La Fuente and A. Flores-Tlacuahuac, "Integrated design and control using a simultaneous mixed-integer dynamic optimization approach," *Industrial and Engineering Chemistry Research*, vol. 48, no. 4, pp. 1933–1943, 2009.

429. X. Li, A. Tomasgard, and P. Barton, "Decomposition strategy for natural gas production network design under uncertainty," in *Proceedings of the IEEE Conference on Decision and Control*, pp. 188–193, 2010.
430. Y. Chen, T. Adams, and P. Barton, "Optimal design and operation of flexible energy polygeneration systems," *Industrial and Engineering Chemistry Research*, vol. 50, no. 8, pp. 4553–4566, 2011.
431. Y. Chen, T. Adams, and P. Barton, "Optimal design and operation of static energy polygeneration systems," *Industrial and Engineering Chemistry Research*, vol. 50, no. 9, pp. 5099–5113, 2011.
432. X. Li and P. Barton, "Optimal design and operation of energy systems under uncertainty," *Journal of Process Control*, vol. 30, pp. 1–9, 2015.
433. A. Ghobeity and A. Mitsos, "Optimal design and operation of a solar energy receiver and storage," *Journal of Solar Energy Engineering, Transactions of the ASME*, vol. 134, no. 3, 2012.
434. L. Zhang, C. Xue, A. Malcolm, K. Kulkarni, and A. Linninger, "Distributed system design under uncertainty," *Industrial and Engineering Chemistry Research*, vol. 45, no. 25, pp. 8352–8360, 2006.
435. H. Li, R. Gani, and S. Jørgensen, "Integration of design and control for energy integrated distillation," *Computer Aided Chemical Engineering*, vol. 14, no. C, pp. 449–454, 2003.
436. E. Ramírez and R. Gani, "Design and control structure integration from a model-based methodology for reaction-separation with recycle systems," *Computer Aided Chemical Engineering*, vol. 20, no. C, pp. 1519–1524, 2005.
437. M. Hamid, G. Sin, and R. Gani, "Integration of process design and controller design for chemical processes using model-based methodology," *Computers and Chemical Engineering*, vol. 34, no. 5, pp. 683–699, 2010.
438. D. Georgis, S. Jogwar, A. Almansoori, and P. Daoutidis, "Design and control of energy integrated sofc systems for in situ hydrogen production and power generation," *Computers and Chemical Engineering*, vol. 35, no. 9, pp. 1691–1704, 2011.
439. H. Lee, L. Koppel, and H. Lim, "Integrated approach to design and control of a class of countercurrent processes," *Industrial and Engineering Chemistry: Process Design and Development*, vol. 11, no. 3, pp. 376–382, 1972.
440. K.-Y. Hsu, Y.-C. Hsiao, and I.-L. Chien, "Design and control of dimethyl carbonate-methanol separation via extractive distillation in the dimethyl carbonate reactive-distillation process," *Industrial and Engineering Chemistry Research*, vol. 49, no. 2, pp. 735–749, 2010.
441. S. Gunasekaran, N. Mancini, and A. Mitsos, "Optimal design and operation of membrane-based oxy-combustion power plants," *Energy*, vol. 70, pp. 338–354, 2014.
442. W. Luyben, "Design and control of distillation columns with intermediate reboilers," *Industrial and Engineering Chemistry Research*, vol. 43, no. 26, pp. 8244–8250, 2004.
443. W. Luyben, "Design and control of a fully heat-integrated pressure-swing azeotropic distillation system," *Industrial and Engineering Chemistry Research*, vol. 47, no. 8, pp. 2681–2685, 2008.
444. W. Luyben, "Design and control of the monoisopropylamine process," *Industrial and Engineering Chemistry Research*, vol. 48, no. 23, pp. 10551–10563, 2009.
445. W. Luyben, "Design and control of a methanol reactor/column process," *Industrial and Engineering Chemistry Research*, vol. 49, no. 13, pp. 6150–6163, 2010.
446. W. Luyben and I.-L. Chien, *Design and Control of Distillation Systems for Separating Azeotropes*. John Wiley & Sons, Ltd, 2010.
447. W. Luyben, "Design and control of the cumene process," *Industrial and Engineering Chemistry Research*, vol. 49, no. 2, pp. 719–734, 2010.
448. W. Luyben, "Design and control of the methoxy-methyl-heptane process," *Industrial and Engineering Chemistry Research*, vol. 49, no. 13, pp. 6164–6175, 2010.

449. W. Luyben, "Design and control of a modified vinyl acetate monomer process," *Industrial and Engineering Chemistry Research*, vol. 50, no. 17, pp. 10136–10147, 2011.
450. W. Luyben, "Design and control of the acetone process via dehydrogenation of 2-propanol," *Industrial and Engineering Chemistry Research*, vol. 50, no. 3, pp. 1206–1218, 2011.
451. W. Luyben, "Design and control of the butyl acetate process," *Industrial and Engineering Chemistry Research*, vol. 50, no. 3, pp. 1247–1263, 2011.
452. W. Luyben, "Design and control of the ethyl benzene process," *AIChE Journal*, vol. 57, no. 3, pp. 655–670, 2011.
453. W. Luyben, "Design and control of the styrene process," *Industrial and Engineering Chemistry Research*, vol. 50, no. 3, pp. 1231–1246, 2011.
454. W. Luyben, "Design and control of a cooled ammonia reactor," in *Plantwide Control: Recent Developments and Applications* (G. Rangaiah and K. V., eds.), ch. 13, pp. 273–292, John Wiley & Sons, Ltd, 2012.
455. W. Luyben, "Design and control of stacked-column distillation systems," *Industrial and Engineering Chemistry Research*, vol. 53, no. 33, pp. 13139–13145, 2014.
456. S. Stefanis, A. Livingston, and E. Pistikopoulos, "Environmental impact considerations in the optimal design and scheduling of batch processes," *Computers and Chemical Engineering*, vol. 21, no. 10, pp. 1073–1094, 1997.
457. B. Gebreslassie, Y. Yao, and F. You, "Design under uncertainty of hydrocarbon biorefinery supply chains: Multiobjective stochastic programming models, decomposition algorithm, and a comparison between cvar and downside risk," *AIChE Journal*, vol. 58, no. 7, pp. 2155–2179, 2012.
458. F. Bernardo, E. Pistikopoulos, and P. Saraiva, "Robustness criteria in process design optimization under uncertainty," *Computers and Chemical Engineering*, vol. 23, no. SUPPL. 1, pp. S459–S462, 1999.
459. F. Bernardo, P. Saraiva, and E. Pistikopoulos, "Inclusion of information costs in process design optimization under uncertainty," *Computers and Chemical Engineering*, vol. 24, no. 2-7, pp. 1695–1701, 2000.
460. F. Bernardo, P. Saraiva, and E. Pistikopoulos, "Process design under uncertainty: Robustness criteria and value of information," *Computer Aided Chemical Engineering*, vol. 16, no. C, pp. 175–208, 2003.
461. G. Bode, R. Schomäcker, K. Hungerbühler, and G. McRae, "Dealing with risk in development projects for chemical products and processes," *Industrial and Engineering Chemistry Research*, vol. 46, no. 23, pp. 7758–7779, 2007.
462. D. Johnson and I. Bogle, "Handling uncertainty in the development and design of chemical processes," *Reliable Computing*, vol. 12, no. 6, pp. 409–426, 2006.
463. S. Skogestad and M. Morari, "Robust performance of decentralized control systems by independent designs," *Automatica*, vol. 25, no. 1, pp. 119–125, 1989.
464. D. Olsen, W. Svrcek, and B. Young, "Plantwide control study of a vinyl acetate monomer process design," *Chemical Engineering Communications*, vol. 192, no. 10-12, pp. 1243–1257, 2005.
465. Y. Arkun and G. Stephanopoulos, "Studies in the synthesis of control structures for chemical processes: Part iv. design of steady-state optimizing control structures for chemical process units," *AIChE Journal*, vol. 26, no. 6, pp. 975–991, 1980.
466. J. Calandranis and G. Stephanopoulos, "A structural approach to the design of control systems in heat exchanger networks," *Computers and Chemical Engineering*, vol. 12, no. 7, pp. 651–669, 1988.
467. K. Ma, H. Valdés-González, and I. Bogle, "Process design in siso systems with input multiplicity using bifurcation analysis and optimisation," *Journal of Process Control*, vol. 20, no. 3, pp. 241–247, 2010.

468. F. Strutzel and I. D. L. Bogle, "Assessing plant design with regard to MPC performance," *Computers & Chemical Engineering*, vol. 94, pp. 180–211, 2016.
469. P. Seferlis and M. Georgiadis, "The integration of process design and control-summary and future directions," *Computer Aided Chemical Engineering*, vol. 17, no. C, pp. 1–9, 2004.
470. P. Vega, R. Lamanna de Rocco, S. Revollar, and M. Francisco, "Integrated design and control of chemical processes - part i: Revision and classification," *Computers and Chemical Engineering*, vol. 71, pp. 602–617, 2014.
471. L. Ricardez-Sandoval, H. Budman, and P. Douglas, "Integration of design and control for chemical processes: A review of the literature and some recent results," *Annual Reviews in Control*, vol. 33, no. 2, pp. 158–171, 2009.
472. L. Ricardez-Sandoval, "Current challenges in the design and control of multiscale systems," *Canadian Journal of Chemical Engineering*, vol. 89, no. 6, pp. 1324–1341, 2011.
473. Z. Yuan, B. Chen, G. Sin, and R. Gani, "State-of-the-art and progress in the optimization-based simultaneous design and control for chemical processes," *AIChE Journal*, vol. 58, no. 6, pp. 1640–1659, 2012.
474. A. Ghobeity and A. Mitsos, "Optimal design and operation of desalination systems: New challenges and recent advances," *Current Opinion in Chemical Engineering*, vol. 6, pp. 61–68, 2014.
475. C. Panos, K. Kouramas, M. Georgiadis, and E. Pistikopoulos, "Dynamic optimization and robust explicit model predictive control of hydrogen storage tank," *Computers and Chemical Engineering*, vol. 34, no. 9, pp. 1341–1347, 2010.
476. B. Chachuat, A. Singer, and P. Barton, "Global methods for dynamic optimization and mixed-integer dynamic optimization," *Industrial & Engineering Chemistry Research*, vol. 45, no. 25, pp. 8373–8392, 2006.
477. A. Mitsos, B. Chachuat, and P. Barton, "Towards global bilevel dynamic optimization," *Journal of Global Optimization*, vol. 45, no. 1, pp. 63–93, 2009. cited By 12.
478. J. Oldenburg and W. Marquardt, "Disjunctive modeling for optimal control of hybrid systems," *Computers & Chemical Engineering*, vol. 32, no. 10, pp. 2346–2364, 2008.
479. Y. Chu and F. You, "Integrated scheduling and dynamic optimization of sequential batch processes with online implementation," *AIChE Journal*, vol. 59, no. 7, pp. 2379–2406, 2013.
480. Y. Chu and F. You, "Integrated scheduling and dynamic optimization of complex batch processes with general network structure using a generalized benders decomposition approach," *Industrial & Engineering Chemistry Research*, vol. 52, no. 23, pp. 7867–7885, 2013.
481. H. Shi, Y. Chu, and F. You, "Novel optimization model and efficient solution method for integrating dynamic optimization with process operations of continuous manufacturing processes," *Industrial & Engineering Chemistry Research*, vol. 54, no. 7, pp. 2167–2187, 2015.
482. H. Shi and F. You, "A novel adaptive surrogate modeling-based algorithm for simultaneous optimization of sequential batch process scheduling and dynamic operations," *AIChE Journal*, vol. 61, no. 12, pp. 4191–4209, 2015.
483. R. Oberdieck, N. A. Diangelakis, S. Avraamidou, and E. N. Pistikopoulos, "On unbounded and binary parameters in multi-parametric programming: applications to mixed-integer bilevel optimization and duality theory," *Journal of Global Optimization*, 2016, in press.
484. R. Jacobs and R. Krishna, "Multiple solutions in reactive distillation for methyl tert-butyl ether synthesis," *Industrial & Engineering Chemistry Research*, vol. 32, no. 8, pp. 1706–1709, 1993.
485. S. Hauan, T. Hertzberg, and K. M. Lien, "Why methyl tert-butyl ether production by reactive distillation may yield multiple solutions," *Industrial & Engineering Chemistry Research*, vol. 34, no. 3, pp. 987–991, 1995.
486. M. Schenk, R. Gani, D. Bogle, and E. N. Pistikopoulos, "A hybrid modelling approach for separation systems involving distillation," *Chemical Engineering Research and Design*, vol. 77, no. 6, pp. 519–534, 1999.

487. G. J. Harmsen, "Reactive distillation: The front-runner of industrial process intensification: A full review of commercial applications, research, scale-up, design and operation," *Chemical Engineering and Processing: Process Intensification*, vol. 46, no. 9, pp. 774–780, 2007.
488. L. A. Smith Jr, "Method for the preparation of methyl tertiary butyl ether," Dec. 18 1990. US Patent 4,978,807.
489. A. Rehfinger and U. Hoffmann, "Kinetics of methyl tertiary butyl ether liquid phase synthesis catalyzed by ion exchange resin – I. intrinsic rate expression in liquid phase activities," *Chemical Engineering Science*, vol. 45, no. 6, pp. 1605–1617, 1990.
490. F. Colombo, L. Cori, L. Dalloro, and P. Delogu, "Equilibrium constant for the methyl tert-butyl ether liquid-phase synthesis using UNIFAC," *Industrial & Engineering Chemistry Fundamentals*, vol. 22, no. 2, pp. 219–223, 1983.
491. Y. Tian and E. N. Pistikopoulos, "Synthesis of operable process intensification systems – Steady-state design with safety and operability considerations," *Industrial & Engineering Chemistry Research*, vol. 58, no. 15, pp. 6049–6068, 2018.
492. M. A. Schenk, *Design of operable reactive distillation columns.* PhD thesis, University of London, 1999.
493. P. Panjwani, M. Schenk, M. Georgiadis, and E. N. Pistikopoulos, "Optimal design and control of a reactive distillation system," *Engineering Optimization*, vol. 37, no. 7, pp. 733–753, 2005.
494. R. Oberdieck, N. A. Diangelakis, and E. N. Pistikopoulos, "Explicit model predictive control: A connected-graph approach," *Automatica*, vol. 76, pp. 103–112, 2017.
495. N. Diangelakis, S. Avraamidou, and E. Pistikopoulos, "Decentralised multi-parametric model predictive control study for domestic chp systems," *Industrial & Engineering Chemistry Research*, vol. 55, no. 12, pp. 3313–3326, 2016.
496. I. Grossmann, "Enterprise-wide optimization: A new frontier in process systems engineering," *AIChE Journal*, vol. 51, no. 7, pp. 1846–1857, 2005.
497. C. A. Floudas, A. M. Niziolek, O. Onel, and L. R. Matthews, "Multi-scale systems engineering for energy and the environment: Challenges and opportunities," *AIChE Journal*, vol. 62, no. 3, pp. 602–623, 2016.
498. I. Harjunkoski, R. Nyström, and A. Horch, "Integration of scheduling and control-Theory or practice?," *Computers and Chemical Engineering*, vol. 33, no. 12, pp. 1909–1918, 2009.
499. A. Flores-Tlacuahuac and I. Grossmann, "An effective mido approach for the simultaneous cyclic scheduling and control of polymer grade transition operations," *Computer Aided Chemical Engineering*, vol. 21, no. C, pp. 1221–1226, 2006.
500. M. A. Gutiérrez-Limón, A. Flores-Tlacuahuac, and I. E. Grossmann, "MINLP formulation for simultaneous planning, scheduling, and control of short-period single-unit processing systems," *Industrial and Engineering Chemistry Research*, vol. 53, no. 38, pp. 14679–14694, 2014.
501. K. Mitra, R. D. Gudi, S. C. Patwardhan, and G. Sardar, "Resiliency issues in integration of scheduling and control," *Industrial and Engineering Chemistry Research*, vol. 49, no. 1, pp. 222–235, 2010.
502. I. Mitrai and P. Daoutidis, "Decomposition of integrated scheduling and dynamic optimization problems using community detection," *Journal of Process Control*, vol. 90, pp. 63–74, 2020.
503. D. Mora-Mariano, M. A. Gutiérrez-Limón, and A. Flores-Tlacuahuac, "A lagrangean decomposition optimization approach for long-term planning, scheduling and control," *Computers & Chemical Engineering*, vol. 135, p. 106713, 2020.
504. C. Chatzidoukas, C. Kiparissides, J. D. Perkins, and E. N. Pistikopoulos, "Optimal grade transition campaign scheduling in a gas-phase polyolefin FBR using mixed integer dynamic optimization," *Computer Aided Chemical Engineering*, vol. 15, no. C, pp. 744–747, 2003.
505. Y. Chu and F. You, "Integration of scheduling and control with online closed-loop implementation: Fast computational strategy and large-scale global optimization algorithm," *Computers and Chemical Engineering*, vol. 47, pp. 248–268, 2012.

506. E. Capón-García, G. Guillén-Gosálbez, and A. Espuña, "Integrating process dynamics within batch process scheduling via mixed-integer dynamic optimization," *Chemical Engineering Science*, vol. 102, pp. 139–150, 2013.
507. J. Park, J. Du, I. Harjunkoski, and M. Baldea, "Integration of scheduling and control using internal coupling models," *Computer Aided Chemical Engineering*, vol. 33, pp. 529–534, 2014.
508. R. H. Nyström, R. Franke, I. Harjunkoski, and A. Kroll, "Production campaign planning including grade transition sequencing and dynamic optimization," vol. 29, pp. 2163–2179, 2005.
509. G. M. Kopanos, M. C. Georgiadis, and E. N. Pistikopoulos, "Energy production planning of a network of micro combined heat and power generators," *Applied Energy*, vol. 102, pp. 1522–1534, 2013.
510. L. S. Dias and M. G. Ierapetritou, "110th anniversary: Integration of scheduling and robust model predictive control," *Industrial & Engineering Chemistry Research*, vol. 59, no. 1, pp. 265–280, 2020.
511. O. Andrés-Martínez and L. A. Ricardez-Sandoval, "A switched system formulation for optimal integration of scheduling and control in multi-product continuous processes," *Journal of Process Control*, vol. 106, pp. 94–109, 2021.
512. N. A. Diangelakis and E. N. Pistikopoulos, "Model-based multi-parametric programming strategies towards the integration of design, control and operational optimization," in *27th European Symposium on Computer-Aided Process Engineering (ESCAPE-27)*, pp. 1867–1872, Elsevier, 2017.
513. S. J. Honkomp, L. Mockus, and G. V. Reklaitis, "A framework for schedule evaluation with processing uncertainty," *Computers and Chemical Engineering*, vol. 23, no. 4-5, pp. 595–609, 1999.
514. X. Lin, S. L. Janak, and C. A. Floudas, "A new robust optimization approach for scheduling under uncertainty: I. Bounded uncertainty," *Computers and Chemical Engineering*, vol. 28, no. 6-7, pp. 1069–1085, 2004.
515. S. L. Janak, X. Lin, and C. A. Floudas, "A new robust optimization approach for scheduling under uncertainty. II. Uncertainty with known probability distribution," *Computers and Chemical Engineering*, vol. 31, no. 3, pp. 171–195, 2007.
516. Y. Chu and F. You, "Model-based integration of control and operations: Overview, challenges, advances, and opportunities," *Computers and Chemical Engineering*, vol. 83, pp. 2–20, 2015.
517. Y. Peng and L. Ricardez-Sandoval, "Integration of planning, scheduling, and control for multi-product chemical systems under preventive maintenance," *IFAC-PapersOnLine*, vol. 54, no. 3, pp. 536–541, 2021. 16th IFAC Symposium on Advanced Control of Chemical Processes ADCHEM 2021.
518. M. Bassett, P. Dave, F. Doyle, G. Kudva, J. Pekny, G. Reklaitis, S. Subrahmanyam, D. Miller, and M. Zentner, "Perspectives on model based integration of process operations," *Computers & Chemical Engineering*, vol. 20, no. 6, pp. 821–844, 1996.
519. C. A. Floudas and X. Lin, "Continuous-time versus discrete-time approaches for scheduling of chemical processes: A review," *Computers and Chemical Engineering*, vol. 28, no. 11, pp. 2109–2129, 2004.
520. A. Caspari, C. Tsay, A. Mhamdi, M. Baldea, and A. Mitsos, "The integration of scheduling and control: Top-down vs. bottom-up," *Journal of Process Control*, vol. 91, pp. 50–62, 2020.
521. N. A. Diangelakis, *Model-based multi-parametric programming strategies towards the integration of design, control and operational optimization*. PhD thesis, Imperial College London, London, United Kingdom, 2017.
522. P. Grieder, M. Kvasnica, M. Baotic, and M. Morari, "Low Complexity Control of Piecewise Affine Systems with Stability Guarantee," *Proceedings of the 2004 American Control Conference, Boston, Massachusetts*, pp. 1196–1201, 2004.

523. P. Daoutidis, J. H. Lee, I. Harjunkoski, S. Skogestad, M. Baldea, and C. Georgakis, "Integrating operations and control : A perspective and roadmap for future research," *Computers and Chemical Engineering*, vol. 115, pp. 179–184, 2018.
524. R. C. Pattison, C. R. Touretzky, T. Johansson, I. Harjunkoski, and M. Baldea, "Optimal Process Operations in Fast-Changing Electricity Markets: Framework for Scheduling with Low-Order Dynamic Models and an Air Separation Application," *Industrial and Engineering Chemistry Research*, vol. 55, no. 16, pp. 4562–4584, 2016.
525. M. T. Kelley, R. C. Pattison, R. Baldick, and M. Baldea, "An efficient milp framework for integrating nonlinear process dynamics and control in optimal production scheduling calculations," *Computers & Chemical Engineering*, vol. 110, pp. 35–52, 2018.
526. J. G. Costandy, T. F. Edgar, and M. Baldea, "A scheduling perspective on the monetary value of improving process control," *Computers and Chemical Engineering*, vol. 112, pp. 121–131, 2018.
527. M. Ellis and P. D. Christofides, "Integrating dynamic economic optimization and model predictive control for optimal operation of nonlinear process systems," *Control Engineering Practice*, vol. 22, pp. 242–251, 2014.
528. M. Ellis and P. D. Christofides, "Real-time economic model predictive control of nonlinear process systems," *AIChE Journal*, vol. 61, no. 2, pp. 555–571, 2015.
529. A. Alanqar, H. Durand, F. Albalawi, and P. D. Christofides, "An economic model predictive control approach to integrated production management and process operation," *AIChE Journal*, vol. 63, no. 6, pp. 1892–1906, 2017.
530. L. D. Beal, D. Petersen, D. Grimsman, S. Warnick, and J. D. Hedengren, "Integrated scheduling and control in discrete-time with dynamic parameters and constraints," *Computers & Chemical Engineering*, vol. 115, pp. 361–376, 2018.
531. S. Liu and J. Liu, "Economic model predictive control with extended horizon," *Automatica*, vol. 73, pp. 180–192, 2016.
532. L. S. Dias and M. G. Ierapetritou, "From process control to supply chain management: An overview of integrated decision making strategies," *Computers & Chemical Engineering*, vol. 106, pp. 826–835, 2017.
533. B. Burnak, J. Katz, N. A. Diangelakis, and E. N. Pistikopoulos, "Integration of design, scheduling, and control of combined heat and power systems: A multiparametric programming based approach," in *Computer Aided Chemical Engineering*, vol. 44, pp. 2203–2208, Elsevier, 2018.
534. G. Towler and R. Sinnott, "Chapter 7 - capital cost estimating," in *Chemical Engineering Design (Second Edition)* (G. Towler and R. Sinnott, eds.), pp. 307–354, Boston: Butterworth-Heinemann, second edition ed., 2013.
535. N. A. Diangelakis, S. Avraamidou, and E. N. Pistikopoulos, "Decentralized Multiparametric Model Predictive Control for Domestic Combined Heat and Power Systems," *Industrial & Engineering Chemistry Research*, vol. 55, no. 12, pp. 3313–3326, 2016.
536. D. Rippin, "Design and operation of multiproduct and multipurpose batch chemical plants. — an analysis of problem structure," *Computers & Chemical Engineering*, vol. 7, no. 4, pp. 463–481, 1983.
537. C. C. Pantelides, "Unified frameworks for optimal process planning and scheduling," in *Proceedings on the second conference on foundations of computer aided operations*, pp. 253–274, Cache Publications New York, 1993.
538. M. Tawarmalani and N. V. Sahinidis, "A polyhedral branch-and-cut approach to global optimization," *Mathematical Programming*, vol. 103, pp. 225–249, 2005.
539. "Economic indicators," *Chemical Engineering*, vol. 125, p. 64, 7 2018.

MIX
Papier aus verantwortungsvollen Quellen
Paper from responsible sources
FSC® C105338

If you have any concerns about our products,
you can contact us on
ProductSafety@springernature.com

In case Publisher is established outside the EU,
the EU authorized representative is:
.Springer Nature Customer Service Center GmbH
Europaplatz 3, 69115 Heidelberg, Germany

Printed by Libri Plureos GmbH
in Hamburg, Germany